Predicting from Data

An Alternative Unit for Representing and Analyzing Two-Variable Data

GLENCOE
Mathematics Replacement Units

GLENCOE

McGraw-Hill

New York, New York
Columbus, Ohio
Mission Hills, California
Peoria, Illinois

Printed in the United States of America.

Send all inquiries to:
Glencoe/McGraw-Hill
936 Eastwind Drive
Westerville, Ohio 43081

ISBN: 0-02-824208-4 (Student Edition)
ISBN: 0-02-824209-2 (Teacher's Annotated Edition)

1 2 3 4 5 6 7 8 9 10 VH/LH-P 03 02 01 00 99 98 97 96 95 94

NEW DIRECTIONS IN THE MATHEMATICS CURRICULUM

Predicting from Data is a replacement unit developed to provide an alternative to the traditional method of presentation of selected topics in Pre-Algebra, Algebra 1, Geometry, and Algebra 2.

The NCTM Board of Directors' Statement on Algebra says,
> "Making algebra count for everyone will take sustained commitment, time and resources on the part of every school district. As a start, it is recommended that local districts–... 3. experiment with replacement units specifically designed to make algebra accessible to a broader student population." (May, 1994 *NCTM News Bulletin*.)

This unit uses data analysis as a context to introduce and connect broadly useful ideas in statistics and algebra. It is organized around multi-day lessons called investigations. Each investigation consists of several related activities designed to be completed by students working together in cooperative groups. The focus of the unit is on the development of mathematical thinking and communication. Students should have access to computers with statistical software and/or calculators capable of producing graphs and lines of best fit.

About the Authors

Christian R. Hirsch is a Professor of Mathematics and Mathematics Education at Western Michigan University, Kalamazoo, Michigan. He received his Ph.D. degree in mathematics education from the University of Iowa. He has had extensive high school and college level mathematics teaching experience. Dr. Hirsch was a member of the NCTM's Commission on Standards for School Mathematics and chairman of its Working Group on Curriculum for Grades 9-12. He is the author of numerous articles in mathematics education journals and is the editor of several NCTM publications, including the *Curriculum and Evaluation Standards for School Mathematics Addenda Series, Grades 9-12*. Dr. Hirsch has served as president of the Michigan Council of Teachers of Mathematics and on the Board of Directors of the School Science and Mathematics Association. He is currently a member of the NCTM Board of Directors.

Arthur F. Coxford is a Professor of Mathematics Education and former Chairman of the Teacher Education Program at the University of Michigan, Ann Arbor, Michigan. He received his Ph.D. in mathematics education from the University of Michigan. He has been involved in mathematics education for over 30 years. Dr. Coxford is active in numerous professional organizations such as the National Council of Teachers of Mathematics, for which he was the editor of the 1988 NCTM Yearbook, *The Ideas of Algebra, K-12*. He was also the general editor of the 1993 and 1994 NCTM Yearbooks. Dr. Coxford has served as the president of the Michigan Council of Teachers of Mathematics and as the president of the School Science and Mathematics Association. He is presently the general editor of the 1995 NCTM Yearbook.

CONSULTANTS

Each of the Consultants read all five investigations. They gave suggestions for improving the Student Edition and the Teaching Suggestions and Strategies in the Teacher's Annotated Edition.

Richie Berman, Ph.D.
Teacher Education Program
University of California
Santa Barbara, California

Linda Bowers
Mathematics Teacher
Alcorn Central High School
Glen, Mississippi

William Collins
Mathematics Teacher
James Lick High School
San Jose, California

David D. Molina, Ph.D.
E. Glenadine Gibb Fellow in
  Mathematics Education &
  Assistant Professor
The University of Texas at Austin
Austin, Texas

Louise Petermann
Mathematics Curriculum Coordinator
Anchorage School District
Anchorage, Alaska

Dianne Pors
Mathematics Curriculum Coordinator
East Side Union High School
San Jose, California

Javier Solerzano
Mathematics Teacher
South El Monte High School
South El Monte, California

Table of Contents

Making Mathematics Accessible to All: First-Year Pilot Teachers

The authors would like to acknowledge the following people who field tested preliminary versions of *Exploring Data* and *Predicting with Data* in the schools indicated and whose experiences supported the development of the Teacher's Annotated Editions.

Ellen Bacon
Bedford High School
Bedford, Michigan

Elizabeth Berg
Dominican High School
Detroit, Michigan

Nancy Birkenhauer
North Branch High
   School
North Branch, Michigan

Peggy Bosworth
Plymouth Canton High
   School
Canton, Michigan

Bruce Buzynski
Ludington High School
Ludington, Michigan

Sandy Clark
Hackett Catholic Central
   High School
Kalamazoo, Michigan

Tom Duffey
Marshall High School
Marshall, Michigan

Lonney Evon
Quincy High School
Quincy, Michigan

Carole Fielek
Edsel Ford High School
Dearborn, Michigan

Stanley Fracker
Michigan Center High
   School
Michigan Center,
   Michigan

Bonnie Frye
Kalamazoo Central High
   School
Kalamazoo, Michigan

Raymond Kossakowski
East Catholic High
   School
Detroit, Michigan

William Leddy
Lamphere High School
Madison Heights,
   Michigan

Dorothy Louden
Gull Lake High School
Richland, Michigan

Michael McClain
Harry S. Truman High
   School
Taylor, Michigan

Diane Molitoris
Regina High School
Harper Woods, Michigan

Rose Martin
Battle Creek Central
   High School
Battle Creek, Michigan

Carol Nieman
Delton-Kellog High
   School
Delton-Kellog, Michigan

Beth Ritsema
Western Michigan
   University
Kalamazoo, Michigan

John Schneider
North Branch High
   School
North Branch, Michigan

Katherine Smiley
Edsel Ford High School
Dearborn, Michigan

Mark Thompson
Dryden High School
Dryden, Michigan

Paul Townsend
W.K. Kellog Middle
   School
Battle Creek, Michigan

William Trombley
Norway High School
Norway, Michigan

Carolyn White
East Catholic High
   School
Detroit, Michigan

To the Student

T he most often asked question in mathematics classes must be "When am I ever going to use this?" One of the major purposes of *Predicting from Data* is to provide you with a positive answer to this question.

There are several characteristics that this unit has that you may have not experienced before. Some of those characteristics are described below.

Investigations   *Predicting from Data* consists of five investigations. Each investigation has one, two, or three related activities. After a class discussion introduces an investigation or activity, you will probably be asked to work cooperatively with other students in small groups as you gather data, look for patterns, and make conjectures.

Projects   A project is a long-term activity that may involve gathering and analyzing data. You will complete some projects with a group, some with a partner, and some as homework.

Portfolio Assessment   These suggest when to select and store some of your completed work in your portfolio.

Share and Summarize   These headings suggest that your class discuss the results found by different groups. This discussion can lead to a better understanding of key ideas. If your point of view is different, be prepared to defend it.

Displaying Paired Data

Mary Molar is the Athletic Director at Lindell High School. She keeps records of the growth and abilities of all the athletes. Every fall she collects data on athletes in each of grades 9-12. The data she records for males includes weight, height, bench-press weight, 40-yard dash time, age, and grade-point average (GPA) in school. For females, she records weight, height, leg-press weight, 40-yard dash times, age, and grade-point average. She keeps all this information in a computer. Her computer will print lists of data for groups of students. Ms. Molar wants to organize and display her data so that relationships used to understand how the athletes are progressing will be more obvious.

Activity 1-1 Looking for Relationships

The data collected at Lindell High School for 20 male and 20 female athletes over four years is displayed in Tables 1-8 on pages 5 and 6. Table 1 contains data for the male students when they were in the ninth grade, Table 2 for the same students when they were in the tenth grade, Table 3 for these students in the eleventh grade, and Table 4 for these students in the twelfth grade. Tables 5-8 contain similar data for 20 female athletes collected over four years.

Before you can use the data effectively, you need to become familiar with the data.

● Group Project

1. a. What is a "bench press"? What is "GPA"? What is a "leg press"?

Share & Summarize

b. What is the unit of measure for the data in each category? Be prepared to explain your selections to the class.

2. a. Prepare box-and-whisker plots of the weights of male and female athletes in each grade.

b. For each gender, how do the distributions of weights vary from grade to grade?

c. In the ninth grade, what is a typical weight of a male athlete? Did you use the mean or the median? Why? How does the typical weight change from grade to grade?

3. a. Prepare box-and-whisker plots of the heights of male and female athletes in each grade.

b. How do the distributions of heights vary for each gender from grade to grade?

c. In the eleventh grade, what is a typical height of a female athlete? How does this compare with the typical height in the other grades?

 **Share & Summarize**

4. a. Do male athletes vary more in height or in weight as ninth graders? Be prepared to explain your reasoning and choice of statistics to the class.

b. Do female athletes vary more in height or in weight as twelfth graders? Be prepared to explain your reasoning and choice of statistics to the class.

Homework Project

5. a. What is the mean bench-press weight for each of the grades?

b. How does it change as male athletes progress from grade to grade?

c. Would your response to the previous question change if you used the median? Explain.

6. a. What is the median leg-press weight for each of the grades?

b. How does it change as female athletes progress from grade to grade?

c. Would your response to the previous question change if you used the mean? Explain.

7. a. Describe the male ninth graders who have the fastest 40-yard dash times.

b. Describe the male ninth graders who bench press the most weight.

c. How do the descriptions in part b differ from those in part a? How are they similar?

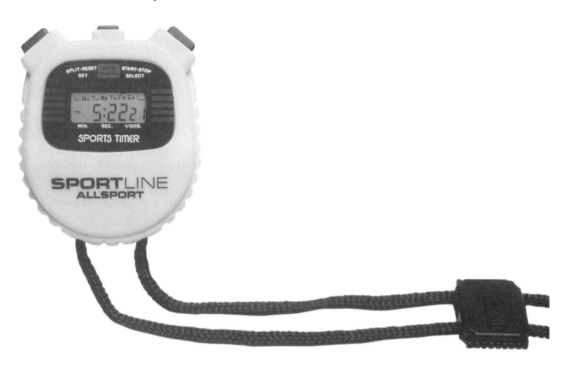

8. Extension Would it be fair to say that younger ninth-grade females run faster or slower than older ninth-grade females? Explain.

Share & Summarize

9. Extension Would it be fair to say that taller ninth-grade females run faster or slower than shorter ninth-grade females? Be prepared to explain your reasoning to the class.

10. Extension If Ms. Molar wanted fast-running players for the soccer team, what sort of ninth grader should she look for?

Table 1: Data for Ninth Grade Male Athletes

Student	Age (mo)	Weight (lb)	Height (in.)	Bench Press (lb)	40-yd time (s)	GPA
1	168	125	65	100	7.2	3.2
2	161	130	67	110	7.5	3.7
3	172	150	70	140	6.1	2.8
4	180	170	73	150	6.0	3.0
5	171	168	69	130	5.4	2.3
6	176	200	73	150	7.6	3.1
7	172	155	73	145	6.1	2.6
8	165	143	65	105	5.9	2.8
9	167	156	69	145	5.6	3.8
10	174	169	70	135	5.5	2.1
11	179	170	75	140	6.3	2.7
12	169	125	69	110	5.5	3.0
13	175	135	62	130	6.0	3.6
14	179	178	71	150	5.7	3.6
15	182	190	77	160	6.3	3.1
16	172	157	71	150	5.9	2.5
17	162	120	67	105	5.9	3.1
18	168	140	65	90	5.7	3.2
19	172	160	68	120	5.3	2.8
20	177	155	69	130	5.8	2.4

Table 2: Data for Tenth Grade Male Athletes

Student	Age (mo)	Weight (lb)	Height (in.)	Bench Press (lb)	40-yd time (s)	GPA
1	180	145	67	120	6.9	3.4
2	173	135	68	115	7.3	3.6
3	184	160	73	150	5.9	2.9
4	192	175	74	160	5.7	3.4
5	183	168	70	145	5.2	2.7
6	188	195	73	165	7.1	3.0
7	184	165	75	160	6.2	2.9
8	177	147	66	130	5.8	2.9
9	179	160	70	155	5.4	3.6
10	186	171	71	140	5.3	2.6
11	191	173	77	165	6.0	2.7
12	181	130	70	125	5.4	3.2
13	187	135	62	130	5.8	3.6
14	191	180	72	165	5.6	3.4
15	194	195	77	175	6.1	2.6
16	184	157	72	160	5.1	2.9
17	174	130	69	115	5.5	3.5
18	180	150	66	110	5.7	3.0
19	184	160	69	140	5.2	2.5
20	189	158	70	150	5.8	2.6

Table 3: Data for Eleventh Grade Male Athletes

Student	Age (mo)	Weight (lb)	Height (in.)	Bench Press (lb)	40-yd time (s)	GPA
1	192	165	71	150	6.2	3.6
2	185	145	68	120	7.1	3.8
3	196	171	74	165	5.6	2.5
4	204	180	74	170	5.5	3.6
5	195	170	72	160	5.1	2.9
6	200	196	75	180	6.8	3.3
7	196	170	75	170	6.1	2.5
8	189	155	70	150	5.5	2.8
9	191	165	70	160	5.4	3.6
10	198	176	73	150	5.2	2.8
11	203	179	77	190	5.9	2.9
12	193	140	72	140	5.3	3.4
13	199	145	66	140	5.7	3.6
14	203	185	73	180	5.4	3.2
15	206	200	77	200	5.9	2.8
16	196	160	73	170	5.0	2.8
17	186	140	70	130	5.4	3.5
18	192	155	67	130	5.6	3.1
19	196	165	70	150	5.2	2.7
20	201	169	70	160	5.4	2.9

Table 4: Data for Twelfth Grade Male Athletes

Student	Age (mo)	Weight (lb)	Height (in.)	Bench Press (lb)	40-yd time (s)	GPA
1	204	175	72	160	5.9	3.4
2	197	167	71	160	6.8	3.9
3	208	180	74	170	5.8	2.9
4	216	183	74	200	5.4	3.4
5	207	175	72	180	5.0	3.0
6	212	200	75	210	6.6	3.5
7	208	172	75	175	6.0	2.6
8	201	155	71	160	5.6	2.9
9	203	170	70	170	5.3	3.5
10	210	173	73	170	5.1	2.9
11	215	180	77	200	5.5	3.0
12	205	150	72	145	5.4	3.5
13	211	150	66	130	5.8	3.8
14	215	187	74	190	5.4	3.2
15	218	200	77	220	5.6	2.9
16	208	165	73	190	5.0	2.6
17	198	150	71	150	5.3	3.5
18	204	165	69	160	5.3	3.3
19	208	170	72	170	5.2	2.5
20	213	172	71	165	5.6	2.6

Table 5: Data for Ninth Grade Female Athletes

Student	Age (mo)	Weight (lb)	Height (in.)	Bench Press (lb)	40-yd time (s)	GPA
1	172	127	68	150	6.7	3.2
2	165	104	58	120	6.2	2.7
3	180	130	66	170	6.0	3.6
4	167	107	59	135	6.8	2.5
5	173	111	60	140	5.8	2.3
6	179	125	67	180	5.7	3.7
7	168	109	59	150	5.6	3.5
8	172	110	62	165	5.5	2.6
9	170	132	67	185	6.0	3.1
10	165	90	59	140	5.8	2.4
11	167	97	63	130	5.7	2.7
12	178	115	61	155	5.5	2.9
13	180	130	65	195	5.7	3.3
14	166	102	60	120	6.2	3.7
15	172	117	64	145	6.0	2.3
16	176	111	63	160	5.9	3.5
17	171	135	68	170	5.9	2.8
18	165	95	61	140	5.4	3.1
19	179	121	67	155	5.5	2.1
20	168	103	62	150	5.3	3.4

Table 6: Data for Tenth Grade Female Athletes

Student	Age (mo)	Weight (lb)	Height (in.)	Bench Press (lb)	40-yd time (s)	GPA
1	184	130	68	165	6.5	3.3
2	179	110	59	140	6.1	2.5
3	192	132	66	185	5.8	3.5
4	189	113	59	150	6.3	2.7
5	185	114	61	150	5.7	2.2
6	191	125	67	190	5.4	3.6
7	180	115	60	165	5.6	3.2
8	184	110	64	170	5.3	2.9
9	182	130	67	195	5.8	3.4
10	177	95	60	160	5.6	2.5
11	179	102	64	140	5.5	2.4
12	190	117	61	155	5.2	2.8
13	192	135	65	200	5.6	3.1
14	178	108	62	125	6.1	3.8
15	184	115	65	160	5.9	2.4
16	188	121	64	165	5.9	3.1
17	183	135	68	180	5.7	2.9
18	177	100	63	150	5.2	3.1
19	191	124	67	165	5.4	2.2
20	180	108	63	155	5.1	3.5

Table 7: Data for Eleventh Grade Female Athletes

Student	Age (mo)	Weight (lb)	Height (in.)	Bench Press (lb)	40-yd time (s)	GPA
1	196	135	68	175	6.4	3.0
2	191	112	60	150	6.0	2.3
3	204	139	66	195	5.8	3.6
4	201	115	60	155	6.1	2.9
5	197	112	61	150	5.5	2.4
6	203	125	67	200	5.4	3.5
7	192	120	60	175	5.5	3.4
8	196	124	64	180	5.2	3.0
9	194	132	67	210	5.8	3.4
10	189	98	61	160	5.4	2.7
11	191	104	64	150	5.3	2.3
12	202	115	61	165	5.2	2.9
13	204	135	65	215	5.6	3.2
14	190	110	62	135	5.8	3.8
15	196	115	65	160	5.8	2.5
16	200	117	64	170	5.9	3.2
17	195	138	68	185	5.6	3.0
18	189	102	63	160	5.1	3.0
19	203	125	68	180	5.4	2.3
20	192	110	63	170	5.1	3.4

Table 8: Data for Twelfth Grade Female Athletes

Student	Age (mo)	Weight (lb)	Height (in.)	Bench Press (lb)	40-yd time (s)	GPA
1	208	139	68	190	6.3	2.9
2	203	115	60	160	5.9	2.4
3	216	140	66	205	5.8	3.6
4	213	117	60	160	6.1	3.0
5	209	110	61	165	5.4	2.3
6	215	129	67	210	5.3	3.3
7	204	118	61	180	5.5	3.4
8	208	128	64	190	5.1	3.0
9	206	135	67	225	5.7	3.3
10	201	102	61	170	5.3	2.7
11	203	110	65	150	5.3	2.3
12	214	118	62	170	5.1	3.0
13	216	142	65	220	5.6	3.1
14	202	112	62	140	5.7	3.8
15	208	119	65	170	5.8	2.5
16	212	114	64	175	5.8	3.1
17	207	135	69	200	5.5	3.0
18	201	105	64	160	5.0	2.9
19	215	131	68	185	5.3	2.3
20	204	116	64	180	4.9	3.4

Activity 1-2   Scatter Plots

Materials

 tape measure

 graph paper

 cylindrical objects

 string

Often it is not easy to look at number pairs and "see" whether or not they are associated in some way. Graphing the pairs of data on a coordinate system is one way to organize the data so that associations are easier to "see." Such a graph is called a **scatter plot**.

Group Project 1

1. Below is a procedure to construct a scatter plot for "height" and "armspan" for your class.

Name	Height (in.)	Armspan (in.)

a. Measure, in inches, the height and armspan of each person in your group. Record your data in a table like the one shown at the right.

Share & Summarize

b. Extend your table and combine your data with that of the other groups.

c. On a piece of graph paper, draw a large ∟ to represent the horizontal and vertical axes, and label the axes. For these data, write *Height (in.)* along the horizontal axis and *Armspan (in.)* along the vertical axis.

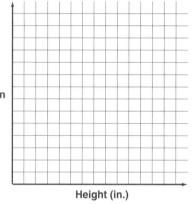

d. Number the tick marks on each axis to make a scale appropriate for the data given.

e. Plot each point (height, armspan) using the pairs of values taken from the class data you collected in part b above as coordinates.

Your graph is a scatter plot of *armspan versus height.*

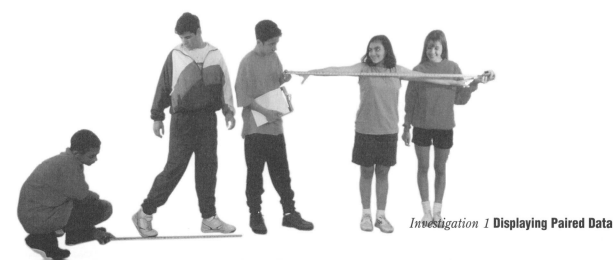

2. Does it appear that there is an association between the armspan and height of students in your class? That is, is an increase in one accompanied by an increase in the other, or is an increase in one accompanied by a decrease in the other? If either case exists, we say they are **associated**. Describe any patterns you see in the scatter plot of these data.

3. Are there any clusters of points, that is, a grouping of points close together and somewhat separated from the other points? If there are, describe the characteristics of each cluster.

4. a. Use this scatter plot to estimate the armspan of a classmate whose height is 65 inches.

 b. Estimate the height of a classmate whose armspan is 68 inches.

 c. Explain how you made your estimates using the scatter plot.

5. Describe how you would make a scatter plot of *height versus armspan*.

The scatter plot for your armspan versus height data depicts a **positive association** in the data. It is positive because the points of the scatter plot rise as the values on the horizontal axis increase. An increase in one variable is accompanied by an increase in the other.

6. a. Collect from each classmate the distance, in inches, from the ceiling to the tip of the fingers reaching toward the ceiling.

Share & Summarize

 b. Make a scatter plot of the distance to the ceiling from fingertip versus height. Label the horizontal axis *Height (in.)* and the vertical axis *Distance to ceiling from fingertip (in.)*. Put a scale on each so that each point can be graphed and so that the points are spread out along the horizontal axis. Describe how you chose the scale for each axis. Be prepared to share your work with the class.

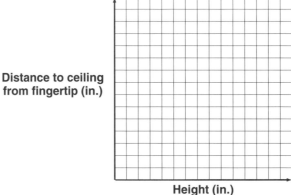

7. Do you think the two variables, height and distance to ceiling from fingertips, are associated? Is the association positive? Give your reasoning.

8. Describe the association as you see it.

9. a. Are there any clusters of points? If so, describe the characteristics of the students in these clusters.

   b. Are there any outliers, that is, individual points that are substantially removed from the rest?

10. a. Predict, using the scatter plot, the distance to the ceiling from the fingertip of a student 65 inches tall.

   b. Predict the height of a student whose distance to the ceiling from fingertips is 30 inches. Explain why you chose this value.

Your scatter plot of the distance to ceiling from fingertip versus height data illustrates a **negative association** since as the height of students increases, the distance to the ceiling decreases. In general, the points on the scatter plot get closer to the horizontal axis as the values on the horizontal axis increase.

Graphing Calculator Activity

You can learn how to use a graphing calculator to make a scatter plot in Activity 1 on page 58.

Partner Project

11. a. Use computer software or a graphing calculator to make a scatter plot of *Shoe Length (in.) versus Month of Birth*. Recall that the vertical axis is labeled "Shoe Length (in.)" while "Month of Birth" is used to label the horizontal axis.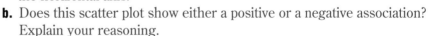

   b. Does this scatter plot show either a positive or a negative association? Explain your reasoning.

   c. How would you label the axes for a scatter plot showing *Month of Birth versus Shoe Length*?

12. Could you predict the month of birth of a person on the basis of his or her shoe length? Explain.

13. Describe the appearance of the points in the scatter plot.

The data for shoe length and month of birth show no consistent association. The month of birth of a person may increase or decrease with increasing shoe length. The points are scattered all over the graph. We say that there is **no association**. You cannot make reliable predictions from such data.

14. a. Make a table to record the measurements of the diameter and circumference of the objects you are provided. Wrap a string around each cylindrical object and measure the string to determine the circumference. Then place the object on the tape measure and find the diameter as accurately as possible. Record the measurements in your table. Combine your measurements with those of your classmates.

b. Use technology to make a scatter plot of *circumference versus diameter*. Which axis should be labeled circumference?

c. Does the data show any association? Is the association positive, negative, or non-existent?

d. Describe the scatter plot.

The *circumference versus diameter* scatter plot showed points that appear to lie in a line. A scatter plot with all the points nearly on a line shows a very strong association. It could be positive or it could be negative, but in either case, it is very strong.

● Group Project 2

15. Sketch three scatter plots each having 15 data points. The first should show a positive association, the second a negative association, and the third little or no association. Compare your plots with that of your group members. Are they similar? Which seems to show the strongest association?

16. Extension Suppose *A* and *B* represent two variables for which you have collected data. If a scatter plot of the *A versus B* data shows a positive association, what can you conclude about a scatter plot of *B versus A*? Explain your reasoning.

17. Extension Within your group, determine the names of ten rock groups you consider most popular. Each member should make a list of these groups. Each member should then choose a classmate from another group whom they think has similar or dissimilar taste in rock groups. Ask that person to rank your ten groups by placing a 1 after the one liked best, a 2 after the next, and so on. While this is happening, you do the same using your ten groups. Summarize the data in a scatter plot with your ranking as the first coordinate and your classmate's ranking as the second coordinate. Are the two ratings associated? Would you conclude that the two of you have similar or dissimilar taste in rock groups? Explain.

 Journal

18. a. Journal Entry   What is a scatter plot?

b. Journal Entry   Illustrate a scatter plot showing positive association, one showing negative association, and one showing little or no association.

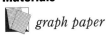
Activity 1-3 Using Scatter Plots to Analyze Data

Computer software and/or graphing calculators simplify the task of constructing scatter plots so you can focus on analyzing the graph, identifying patterns, and communicating your thinking about the data. Recall the Lindell High School data on pages 5 and 6 for males and for females. Choose either males or females for your work in this activity. If a printer is available, print a copy of your scatter plots for future reference and to share with others.

● GROUP PROJECT

1. **a.** Construct a scatter plot for the *bench-press weight* or *leg-press weight* versus *body weight* for Lindell High School ninth graders.

   b. Does it appear that there is an association between the data sets? If so, is it positive or negative?

   c. Does the association appear to be strong, moderate, or weak? Explain your answer.

2. Are there any clusters of points? If so, describe the characteristics of each cluster.

3. Use this scatter plot to estimate the bench-press weight that a 170-pound ninth grade male could lift or the leg-press weight a 115-pound ninth grade female could lift. How did you use the scatter plot to make your estimate? Be specific.

4. Estimate the body weight of a ninth grade male who bench presses 170 pounds or the body weight of a ninth grade female who leg presses 145 pounds using your scatter plot. How did you use the scatter plot to make this estimate? What decisions did you have to make?

Share & Summarize

5. Does your scatter plot show all the points? If not, what points represent more than one item of data? How could you indicate on the scatter plot that a point represented two or more occurrences of an ordered pair in the data? Compare your method with that of a classmate. Which do you prefer? Why? Be prepared to share your methods with the class.

6. Construct a scatter plot for the body weight and 40-yard dash data for eleventh graders included in Table 3 or Table 7 of the Lindell H.S. data. Mark any double or triple points in the manner you chose in Exercise 5.

7. Do you think the two variables, body weight and speed in the 40-yard dash, are associated? Is the association positive or negative? Give your reasoning.

8. Are there any clusters of points? If so, describe the characteristics of the students represented by these clusters. Are there any outliers? If so, are they outlying on one or on both variables?

9. Does it make sense to ask for a predicted 40-yard dash time for a tenth grade male weighing 165 pounds or a tenth grade female weighing 115 pounds? If so, do it and explain. If not, explain your reasoning.

10. **a.** Refer to the height and GPA data for seniors included in Table 4 or in Table 8. Would you guess that these variables are associated or not?

    b. Prepare a scatter plot for GPA versus height for the twelfth grade males or females.

    c. Does your scatter plot confirm or disprove your guess? If the plot is associated, describe the association.

    d. Use your scatter plot to predict the GPA of a student who is 68 inches tall. Would you have great confidence in your prediction? Explain.

11. **a.** Construct a scatter plot for the twelfth grade age data versus the ninth grade age data. Use Tables 1 and 4 or Tables 5 and 8.

    b. Are these two sets of data positively or negatively associated? Explain your position.

    c. Is the association strong or weak? Explain.

    d. If you predicted the age of a twelfth grader on the basis of his or her age as a ninth grader, would you have great or little confidence in your prediction? Explain.

    e. What are the characteristics of this graph? Where do all the points in the scatter plot lie?

 Homework Project

Share & Summarize

12. Investigate the association of several other sets of data from Tables 1-4 or Tables 5-8. Choose pairs that you think will show association and other pairs that you think would not show association. Were your hunches correct for these data? Be prepared to share your findings with the class.

13. **Journal Entry** Use the data in Tables 1-4 or in Tables 5-8 to respond to the following questions.

   a. Is speed as a twelfth grader associated with speed as a ninth grader?

   b. Is speed as a twelfth grader associated with strength as a ninth grader (bench press or leg press)?

   c. Is strength as a twelfth grader associated with age as a ninth grader?

   d. Is GPA as a twelfth grader associated with age as a ninth grader?

   e. Make up a similar question. Construct a scatter plot and then use the graph to respond to the question.

14. **Extension** Suppose all the points graphed in a scatter plot lie on a line. If the association is positive, what could the line look like? If it is negative, what could the line look like? Tell why you chose the lines you did.

15. **Extension** If **perfect association** of two sets of data has all the points falling on a line, how could you estimate the strength of an association if all the points did not fall on a line?

Activity 1-4 How Strong is an Association?

   The three scatter plots below all show positive association because the points rise from left to right. The scatter plot on the left shows the strongest positive association because it can be enclosed in a narrow oval loop. We say the data represented in the graph has a **correlation** that is positive and large. The next scatter plot can be enclosed in a loop, but it is wider than the first. Thus the correlation is smaller but still is positive. The final scatter plot has a small positive correlation because the loop is quite fat. In general, the correlation is greater when the points of the scatter plot make a long thin loop.

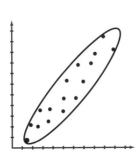 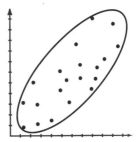 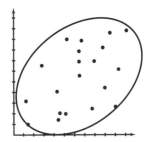

Partner Project

1. The three scatter plots at the top of the next page show negative associations. Why are the associations negative?

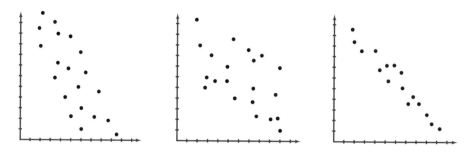

2. Visualize each scatter plot above enclosed with a loop similar to those shown for the variables in the scatter plots on the previous page.

   a. Which scatter plot shows the strongest association? Statisticians say this represents data having a large negative correlation.

   b. Which scatter plot shows the weakest association or smallest negative correlation? Explain your reasoning.

3. If the loop you need to enclose a scatter plot is nearly circular, then the association between the variables is low, and the correlation is nearly zero. Which of the scatter plots below appear to have a very small or weak correlation?

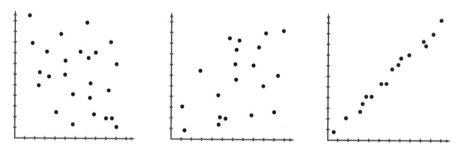

**Share &
Summarize**

4. Correlations are usually reported as numbers between −1 and 1. These are called **correlation coefficients**. A 1 represents perfect positive association, −1 represents perfect negative association, and 0 represents lack of association. Correlation coefficients between 0 and 1 represent positive associations. The stronger the association, the closer the number is to 1. Describe what correlation coefficients between 0 and −1 would mean for the data. Be prepared to share your reasoning with the class.

5. Use software to make a scatter plot for each pair of variables. Print a copy of each scatter plot.

   a. Tenth grade: bench press or leg press versus weight

   b. Twelfth grade: time in 40-yard dash versus weight

   c. Eleventh grade: GPA versus height

   d. Age in grade 12 versus age in grade 10

6. Put loops around each scatter plot in Exercise 5. Estimate what you think the correlation coefficients might be and write them below the appropriate graphs.

Graphing Calculator Activity

You can learn how to use a graphing calculator to find the correlation coefficient in Activity 2, on page 59.

7. **Extension** Graphing calculators and some computer software that display scatter plots will also calculate the correlation coefficient. Use the technology to find the correlation coefficients in Exercise 5 and compare your estimates in Exercise 6 with these values.

8. **Extension** Copy and complete the table of movie preferences listed below. Write a 1 for the movie you like the best, 2 for the second best, and so on to 12 for the least liked. Make a scatter plot of the data. Does there seem to be an association between you and your partner's rankings? Explain.

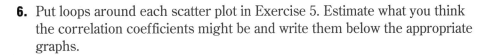

Name of Movie	My Ranking	Partner's Ranking
Home Alone		
Sister Act		
Batman Returns		
Free Willy		
Flintstones		
The Fugitive		
The Beverly Hillbillies		
The Addams Family		
Ace Ventura		
Wayne's World		
A League of Their Own		
Jurassic Park		

Portfolio Assessment

A portfolio is repesentative samples of your work, collected over a period of time. Begin your portfolio by selecting an item that shows something new you learned in this investigation.

9. **Extension** When ranked data, such as that in the table above, have no ties, the correlation coefficient is the same as that of unranked data. The correlation coefficient is called the **Spearman Rank Order Correlation** and may be found using the following formula.

$$r = 1 - \frac{6(\text{sum of the squares of the differences in the rankings})}{n(n^2 - 1)}$$

In this formula, n represents the number of rankings. Use the data in the table of movie preferences to calculate the Spearman Rank Order Correlation.

Journal

10. **Journal Entry** Write a paragraph or two describing what you have learned in this investigation. Include ideas of how you might use this knowledge outside of mathematics class.

Line Fitting

Ms. Molar, the Athletic Director at Lindell High School, is uncertain about how she can use the data about her athletes. She can now see, using scatter plots, that certain characteristics seem to be associated with others while other pairs seem to show little association. She would like to use data to set goals for her athletes – goals the ninth graders can strive for as they go through school. But when she looks at the scatter plots, she sees a lot of variation in twelfth grade performance given similar ninth grade characteristics. When she talks to the ninth graders, what advice can she give regarding each person's goals for running and lifting weights in the future grades?

What Ms. Molar needs is a way to summarize a scatter plot that lets her predict values for a variable on the basis of the values of another variable. A line can summarize a scatter plot that shows either positive or negative correlation. A **summarizing line** follows the general pattern of the scatter plot and goes through the "middle" of the scatter plot. This means that there are about as many points above the line as below it and that the line contains some of the points. Using a line to summarize a scatter plot is called **fitting a line to the data**.

Activity 2-1 Fitting Lines Visually

The 100-meter dash has been run in the Olympics since 1896. The men's times for each of the years through 1988 are given in the table below.

Materials

 graph paper

string

ruler

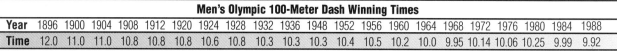

Men's Olympic 100-Meter Dash Winning Times																						
Year	1896	1900	1904	1908	1912	1920	1924	1928	1932	1936	1948	1952	1956	1960	1964	1968	1972	1976	1980	1984	1988
Time	12.0	11.0	11.0	10.8	10.8	10.8	10.6	10.8	10.3	10.3	10.3	10.4	10.5	10.2	10.0	9.95	10.14	10.06	10.25	9.99	9.92

Source: The World Almanac and Book of Facts 1994

PARTNER PROJECT

1. a. Why are there no times for the early 1940s?

   b. Why are the times given in hundredths beginning in 1972?

c. Study the table. Are the winning times associated with the year of the race? If so, describe how they are associated.

2. a. Construct a scatter plot of winning time versus year of race. (Use computer software or a graphing calculator only if they can print a copy of the scatter plot.) What variable does the horizontal axis represent? What variable does the vertical axis represent? (Make sure you choose scales on the axes so that the points spread out across the first quadrant.)

b. Do the data appear to be associated in a linear fashion? If so, is the association positive or negative?

3. a. Use the scatter plot to estimate the winning 100-meter dash time if the Olympics had been held in 1926.

b. Estimate the winning time if the Olympics had been held in 1958.

c. If you were to train in the 1930's for the 100-meter race, what time would allow you to be competitive?

d. What was the best Olympic 100-meter dash time in the 20th century?

e. What was the slowest winning time in the 20th century?

4. a. Place a piece of string on the scatter plot you created in Exercise 2 to visually estimate a line passing through the middle of the scatter plot.

b. Now use a ruler to draw this line. This is your **visually-fit line**.

c. Use this line to estimate the probable winning time if a race had been run in 1926 and in 1958.

d. Compare these estimates with those you made in Exercise 3. Are they close? In which case were the estimates more easily made? Why?

5. The line you drew in Exercise 4 is a **mathematical model** of the data in the table and the scatter plot. It can be used to predict one value of a pair of data when given the other value.

a. Use your line to predict the winning times of the races not run in 1940 and 1944.

b. About what year was a time of 10.5 seconds expected to occur?

c. About what year was a time of 10.0 seconds expected to occur?

d. Use your line to predict the winning time in 1992. Consult an almanac and find out what the winning time was in 1992. Is your prediction close to the winning time for 1992? How close?

6. a. Identify periods of time when the winning times were better than those predicted by your line.

b. Identify periods of time when winning times were worse than those predicted by the line

7. a. Compare your scatter plot and summarizing line with that of your neighbor. Do they differ? Your response should consider mathematical characteristics such as closeness of the points to the line, number of points on the line, the tilt of the line being similar to the tilt of the scatter plot, and so on. Be prepared to share your findings with the class.

b. How different are the values predicted in Exercise 5?

c. How different are the periods identified in Exercise 6?

8. If you extend the line you drew in Exercise 4 in both directions, does it make sense to use it to predict the winning time for a race to be run in the year 2000? in 2020? in 1860? Explain your reasoning.

HOMEWORK PROJECT

9. Examine the data in the table below. Does it appear that the median age of marriage for men and women is related? Why?

10. a. Use computer software or a graphing calculator to construct a scatter plot of male age versus female age. Print a copy of the scatter plot. Do female and male ages at marriage appear to be related?

Median Age at First Marriage

Year	Females	Males
1900	21.9	25.9
1910	21.6	25.1
1920	21.2	24.6
1930	21.3	24.3
1940	21.5	24.3
1950	20.3	22.8
1960	20.3	22.8
1970	20.8	23.2
1980	22.0	24.7
1990	23.9	26.1

Source: Department of Commerce. Bureau of Census

b. Visually estimate a summarizing line and draw it with a ruler.

11. a. Use the line to estimate the age of a man marrying a woman who is 21 years old; 22.5 years old.

b. Use the line to estimate the age of a woman marrying a man who is 23 years old; 24 years old.

12. Make a scatter plot of median male's age of first marriage versus year. Can you visualize a line summarizing this scatter plot? Draw your summarizing line or explain why it would be inappropriate to do so.

13. If a scatter plot does not show an association between the variables, then no line can be used to summarize the data, and none should be drawn. For each of the six scatter plots below and on the next page, decide whether a line should be used to summarize the data. Trace the axes of each scatter plot on a separate piece of paper and then draw the line you think best fits the data. Then predict several values of each variable by reading corresponding values of the other variable from the line. Be prepared to share your work with the class.

Share & Summarize

a. male ninth grade weight versus height

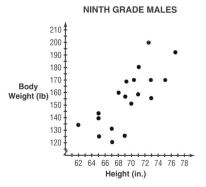

b. female eleventh grade weight versus height

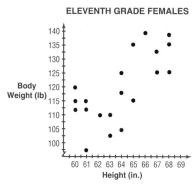

c. male eleventh grade bench press versus time in 40-yard dash

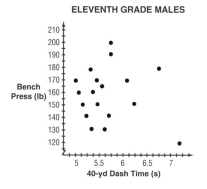

d. female leg press for twelfth graders versus leg press for ninth graders

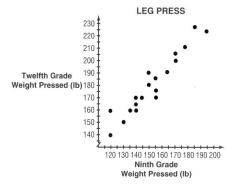

e. female time in 40-yard dash for twelfth graders versus time in 40-yard dash for ninth graders

f. male tenth grade age versus tenth grade GPA

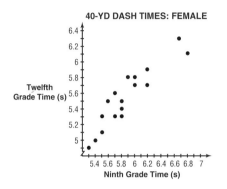

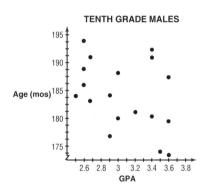

Journal

From this point on, answers for questions asking for summarizing lines or requiring median-fit lines will use the Data Insights program. This program is available from Wings for Learning/Sunburst.

14. Journal Entry

a. Do you think a line is a good mathematical model for the winning Olympic 100-meter dash times? Explain your position and reasoning.

b. If you could summarize these data with a different kind of graph, what would it look like when drawn in the scatter plot? Explain its characteristics and why it has those characteristics.

15. Extension

a. Using computer software, generate other scatter plots for Ms. Molar's data that would be useful to her in advising ninth-grade athletes, both male and female, regarding goals for speed in the 40-yard dash and weight lifted in the bench press or leg press by the time they are in twelfth grade.

b. Summarize each set of data with a line.

c. Use the lines to advise a ninth grader about goals.

Journal

16. Journal Entry

a. Describe how you visually fit a line to a scatter plot.

b. Why are lines fitted to scatter plots?

c. Should every scatter plot be fitted with a line? Explain your position.

Activity 2-2 How Good is Your Fit?

Materials

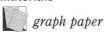
graph paper

ruler

Women have participated in the Olympic high jump since 1928. Below is a table of the heights jumped by women in the high jump event through 1988.

Women's Olympic High Jump Winning Heights

Year	Country	Height
1928	Canada	5 ft 3.000 in.
1932	United States	5 ft 4.250 in.
1936	Hungary	5 ft 3.000 in.
1948	United States	5 ft 6.125 in.
1952	South Africa	5 ft 5.750 in.
1956	United States	5 ft 9.250 in.
1960	Romania	6 ft 0.750 in.
1964	Romania	6 ft 2.750 in.
1968	Czechoslovakia	5 ft 11.750 in.
1972	West Germany	6 ft 3.625 in.
1976	East Germany	6 ft 4.000 in.
1980	Italy	6 ft 5.500 in.
1984	West Germany	6 ft 7.500 in.
1988	United States	6 ft 8.000 in.

Source: Information Please Almanac, 1994

● GROUP PROJECT

1. a. What was the least winning height? When was it a winner?

b. What was the greatest winning height? When was it a winner?

c. What is the mean winning height?

d. What is the median winning height?

2. a. Construct a scatter plot of winning height versus year. Which axis is labeled "Year", and which axis is labeled "Winning Height"?

b. Is the association positive or negative?

c. Is the scatter plot linear in nature?

d. Visually determine a summarizing line and draw it with a ruler.

e. How many points are above the line? How many are below the line? How many are on the line?

3. a. Use a centimeter ruler to measure the vertical distance from each point in the scatter plot to the line you drew in Exercise 2d. This distance is called the **residual** of the linear model of the data. Record each residual.

b. What is the greatest residual?

c. What is the least residual?

d. In terms of inches, what heights do the residuals in parts b and c represent?

Share & Summarize

4. Add the residuals determined in Exercise 3 and find their mean. This **mean residual** is a measure of how well your line fits the data. If the mean is small, then the line is a good fit to the data. A larger mean suggests the line is not a good fit to the data.. How does your mean compare with that of a neighbor? Be prepared to share your reasoning with the class.

Share & Summarize

5. a. Try to summarize the data in the scatter plot in Exercise 2 with another line that you think will improve the fit.
b. Calculate the mean residual for this line and compare it with your original value. Did you improve on your original fit? Be prepared to share your strategies and findings with the class.

6. a. Using the two lines of Exercises 2 and 5, predict the winning women's high jump for 1992 and for 1996.

b. Do the predictions differ? If so, by how much?

c. Which prediction do you have the most confidence in? Explain.

d. Compare your prediction for 1992 with the actual winning height. How close was your prediction?

● HOMEWORK PROJECT

Journal

7. Journal Entry Why do you think it is important to find the summarizing line that fits the data?

FYI
In the 1992 Olympics, the United States women's swim team won a gold medal in 5 of the 15 races.

8. Women began swimming the 100-meter butterfly in the Olympic Games in 1956. The winning times for 1956 and the next eight competitions are given in the table below.

Women's 100-meter Butterfly

Year	Country	Time (s)
1956	United States	71.00
1960	United States	69.50
1964	United States	64.70
1968	Australia	65.50
1972	Japan	63.34
1976	East Germany	60.13
1980	East Germany	60.42
1984	United States	59.26
1988	East Germany	59.00

Source: Information Please Almanac, 1994

Portfolio Assessment
Select one of the assignments from this investigation that you found especially challenging and place it in your portfolio.

a. Use these data to predict the winning time in the 1992 Olympic Games.

b. Are you confident of your prediction? Explain.

8b. Discuss the accuracy of predictions when there are few data points for references.

c. Compare your predicted winning time for 1992 with the actual winning time.

d. Predict the winning time for 1996. Are you confident of your prediction?

9. a. Describe one way used to determine how well a line fits data in a scatter plot.

b. What is a residual?

Journal

10. Journal Entry Write a paragraph or two describing what you have learned in this investigation. Include ideas of types of data that have summarizing lines that are a good fit to the data.

Median-Fit Line

Summarizing the data shown in a scatter plot with a line is not an exact process. As you have seen, several lines may be used to summarize the same data. Some of these lines summarize better than others in the sense that the mean of the residuals is smaller.

Statisticians typically use two methods to draw lines that summarize data. In this investigation, you will use the **median-fit line**. It is easily drawn by hand for small sets of paired data. For larger sets of data, a computer should be used.

Materials

ruler

tracing paper

string

calculator

software

Activity 3-1 Finding and Using Median-Fit Lines

● GROUP PROJECT

1. The table below lists the death rates (deaths per 1000) for United States citizens from 1910 through 1990. What pattern do you see in the data?

United States Deaths per 1,000 Citizens																
Year	1910	1915	1920	1925	1930	1935	1940	1945	1950	1955	1960	1965	1970	1975	1980	1985	1990
Death Rate	14.7	13.2	13.0	11.7	11.3	10.9	10.8	10.6	9.6	9.3	9.5	9.4	9.5	8.8	8.7	8.7	8.6

Source: Information Please Almanac, 1994

2. A computer-generated scatter plot is shown below. Does the data seem to be linear? Represent a visually-fit summarizing line by placing a piece of string on the plot. Be prepared to share your reasoning with the class.

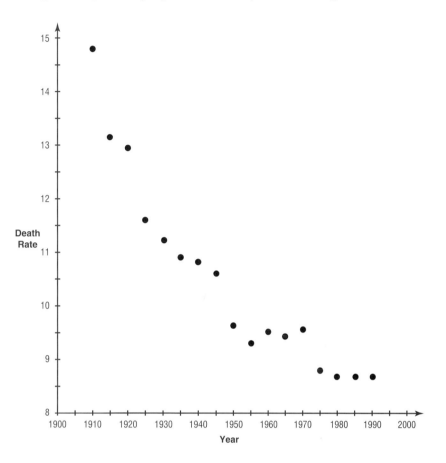

3. The procedure for drawing a median-fit line is described below. Trace the scatter plot above and carry out each of the following instructions on your traced copy.

a. Count the number of data points and divide that total by 3. You will use this number to divide the data points into three sets. If the total is divisible by 3, then the three sets have equal numbers of points. If, however, the total is not divisible by 3, choose the three numbers such that two are equal and larger or smaller than the third. In this example, you have 17 points, which is not divisible by 3, so you choose "thirds" of 6, 6, and 5.

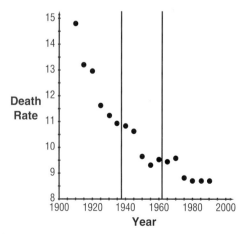

b. Draw two vertical lines on the scatter plot so that the points are divided into three equal sections, or so that the leftmost and rightmost sections are equal and greater than the middle section. For these data you should have sections of 6, 5, and 6, from left to right.

c. The next step is to find the **median point** for each section. The *x*-coordinate of this point is the median of the *x*-coordinates of all the points in the section. The *y*-coordinate is the median of the *y*-coordinates of the points.

Since there are six points in the left section, there are six *x*-values and six *y*-values to use in finding the median point. The median of each set is halfway between the third and fourth values. Find these values.

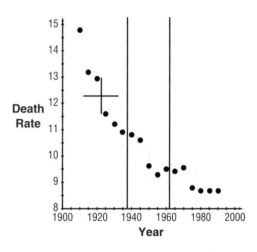

The median point can be found geometrically by placing a ruler horizontally at the bottom of the section and moving it up until it is halfway between the third and fourth points. Draw a short segment here.

Place the ruler vertically at the leftmost edge of the section. Move it to the right until it is halfway between the third and fourth points. Mark a short segment.

The median point of this section is the point where the two short segments intersect.

d. Repeat step c for the other two sections of the scatter plot. In this way you have identified three points that are median points for the three sections of the scatter plot.

e. Place the ruler on the median points in the left and right sections. This determines the slope of the summarizing line. Now slide the ruler one-third the distance to the point in the middle section, keeping the ruler parallel to it original position. Draw this line. The result is the *median-fit line* for these data.

4. Compare your median-fit line with that of your neighbor. Are they nearly the same? If not, how are they different and why are they different?

5. a. On the basis of your median-fit line, predict the death rate in 1900.

   b. What is the predicted death rate in 1950? How far off is the prediction?

6. On the basis of your median-fit line, what is a prediction for the death rate for 1978? for 1989? for 1995? for 2000?

7. a. Is the median-fit line a good model of the death rate data? Explain your reasoning.

   b. Are there portions of the scatter plot that it fits well? not so well? What are they?

   c. What would be the effect of constructing a median-fit line using only 1930-1990 data?

8. a. Compare your visually-fit line to the median-fit line. Are they parallel?

   b. Compute the mean of the residuals for each summarizing line. Which is the better fit?

Share & Summarize

   c. Which summarizing line would you have more confidence in if you were to make predictions for the future? Why? Be prepared to explain your reasoning to the class.

● PARTNER PROJECT

Graphing Calculator Activity

You can learn how to use a graphing calculator to find median-fit lines in Activity 3 on page 60.

9. Use technology to find the median-fit line for the men's 100-meter dash winning times in the Olympic Games from 1896 through 1988 found on page 16. Compare this with your visually-fit line in Exercise 4 on page 17.

10. a. Use technology to find the median-fit line for height versus weight of the ninth-grade males or ninth-grade females found on pages 5 and 6.

   b. If a male weighed 165 pounds or a female weighed 118 pounds, predict their heights.

   c. If a male or female were 66 inches tall, how much would (s)he weigh?

   d. Find the mean of the weights and the mean of the heights for females or males. Plot the point (mean of weights, mean of heights). How close is this point to the median-fit line?

11. a. Use technology to find the median-fit line for height versus weight of eleventh-grade males or eleventh-grade females found on pages 5 and 6.

b. What is the predicted height of a 165-pound male or a 118-pound female?

c. What is the predicted weight of a 6-foot 2-inch male or a 5-foot 4-inch female?

12. a. Use technology to construct the median-fit line for twelfth grade versus ninth grade bench-press or leg-press weight lifting using the data found on pages 5 and 6. Identify all multiple points on your scatter plot.

b. What bench-press or leg-press weight goal would you suggest to a ninth grade athlete who now presses 130 pounds? 160 pounds? 185 pounds?

c. Find the mean of the ninth and twelfth grade bench-press or leg-press weights. Plot the point (mean ninth grade weight, mean twelfth grade weight). How close is this point to the median-fit line?

13. a. Use technology to construct the median-fit line for the times in the 40-yard dash for ninth grade versus twelfth grade males or females found on pages 5 and 6. Identify all multiple points on the scatter plot.

Share & Summarize

b. What time goal, as a senior, would you suggest for a ninth grade athlete who could run the 40-yard dash in 6.6 seconds? in 5.8 seconds? in 6 seconds? Be prepared to share your findings with the class.

14. a. Use technology to construct the median-fit line for the Women's Olympic High Jump Winning Heights on page 21.

b. Predict the winning height for the 1992 and 1996 Olympic Games.

c. If you used this line to predict for the year 2020, what height would you get? Is this a reasonable prediction? Explain.

FYI
Vessels for deep-sea exploration must have complete life-support systems aboard and be able to withstand the enormous pressures of such depths, which range up to 100,000 pounds per square inch.

15. **Extension** The areas and depths of the largest oceans and seas are given in the table below.

Name	Area (1000's of sq. miles)	Average depth (ft)	Greatest depth (ft)
Pacific Ocean	64,000	13,215	36,198
Atlantic Ocean	31,815	12,880	30,246
Indian Ocean	25,300	13,002	24,460
Arctic Ocean	5,440	3,953	18,456
Mediterranean	1,145	4,688	15,197
Caribbean Sea	1,050	8,685	22,788
S. China Sea	895	5,419	16,456
Bering Sea	885	5,075	15,659
Gulf of Mexico	615	4,874	12,425
Okhotsk Sea	614	2,749	12,001
E. China Sea	482	617	9,126
Hudson Bay	476	420	600
Japan Sea	389	4,429	12,276
Andaman Sea	308	2,854	12,392
North Sea	222	308	2,165
Red Sea	169	1,611	7,254
Baltic Sea	163	180	1,380

Source: Information Please Almanac, 1994

a. Locate each of these oceans and seas on a globe or map.

b. Where are the deepest areas located?

c. Use a statistical software package to construct scatter plots for average depth versus area, greatest depth versus area, and greatest depth versus average depth.

d. Visually examine each of these scatter plots. Are they difficult to interpret? Are there clusters of points that are "outliers" from the rest? If so, identify them in each scatter plot.

e. Eliminate any outlier points from each scatter plot and reconstruct. Are relationships more easily observed? Construct the median-fit lines for each scatter plot, if it is appropriate to do so.

f. Determine an equation describing each line. You may use whatever means you need to find the equation. Use each equation to predict the average depth and the maximum depth of a hypothetical sea with an area of 350,000 square miles, 700,000 square miles and 2,000,000 square miles.

16. Extension

a. How could you use means to construct a summarizing line to use in place of the median-fit lines? Describe the procedure to use and illustrate it for the United States Birth Rate data given in the table. The rates are per 1000 people living in the United States.

Portfolio Assessment

Select an item from this investigation that you feel shows your best work and place it in your portfolio. Explain why you selected it.

Birth Rates in the U.S.

Year	Rate
1910	30.1
1915	29.5
1920	27.7
1925	25.1
1930	21.3
1935	18.7
1940	19.4
1945	20.4
1950	24.1
1955	25.0
1960	23.7
1965	19.4
1970	18.4
1975	14.8
1980	15.9
1985	15.8

Source: Information Please Almanac, 1991

b. Historically, what explanation is there for the rates in the 1930's and the rate in the 1950's? According to your linear model, what was the expected rate in those periods?

Linear Patterns

In the last three investigations, you explored methods that would be helpful to Ms. Molar in making sense of student data at Lindell High School. These methods included visual displays and summarizing lines. A summarizing line made it easy to make a reasonable prediction of the future, but we could not be sure the prediction was exactly right. Was your prediction for the 1992 100-meter dash time exactly the same as the winning time? Some situations with which you are familiar give rise to patterns for which a line can be fitted exactly. As a result, predictions can be exact also.

Activity 4-1 Representing Linear Patterns

Materials

graph paper

ruler

calculator

● PARTNER PROJECT

Mega-Hit Video charges $3.00 to rent a video tape for a day and $9.00 to rent a VCR for a day. Juanita and Cedric are planning weekend video parties. Juanita will feature horror movies, and Cedric will show all action tapes.

1. Copy and complete the table of rental costs of the tapes for Juanita shown below.

Number of Tapes (x)	0	1	2	3	4	5	6	7	8
Cost of tapes (in Dollars) (y)	0	3							

2. a. On a sheet of graph paper, make a scatter plot for cost of tapes versus number of tapes.

b. Do the points appear to have a pattern? If so, describe it.

c. Draw the median-fit line for these points.

d. How good is the fit of the median-fit line?

3. a. What are the coordinates of the point where the line and the x-axis intersect?

b. What are the coordinates of the point where the line and the y-axis intersect?

c. Can the line intersect either axis in more than one point? Explain.

4. a. What would be the rental cost if Juanita rented 10 tapes for a day?

b. What would be the rental cost if Juanita rented 50 tapes?

c. Would it make sense for her to rent 50 tapes? Explain.

d. What phrase describes the rental cost if Juanita rented x tapes?

5. a. Write, in words, a rule that seems to describe the relation between the number of tapes rented, x, and the total cost of the rental, y.

b. Compare your description with that of a neighbor. Are they identical? similar? Do they say the same thing?

c. Work together to produce a rule in words you both think is accurate.

d. Translate your rule into an equation, where y represents the cost and x represents the number of tapes rented.

6. a. Use your rule to find the cost of renting 15 tapes.

b. How much would it cost you to rent 26 tapes?

Share & Summarize

c. How many tapes could you rent if you had $25 to spend? Be prepared to share your findings with the class.

● GROUP PROJECT

7. a. Cedric does not have a VCR, so he must rent one. How much rental will Cedric pay to rent a VCR alone for one day? one tape and a VCR?

b. Copy and complete the table for Cedric's cost to rent movies and one VCR for one day.

VCR and Number of Tapes (x)	0	1	2	3	4	5	6	7	8
Cost of VCR and Tapes (in Dollars) (y)		12							

8. **a.** On a sheet of graph paper, make a scatter plot for cost (y) versus VCR and number of tapes (x).

   b. Describe any pattern(s) you see in the points.

   c. Draw the median-fit line for these points.

   d. How good is the fit of the median-fit line to the points?

   e. What is the correlation of these data points?

9. **a.** What are the coordinates of the point where the median-fit line and the y-axis intersect?

   b. What does the value of the first coordinate represent? What does the value of the second coordinate represent?

10. **a.** Does the median-fit line intersect the horizontal or x-axis?

    b. If you extend the axis and the median-fit line, do they intersect?

    c. What are the coordinates of the point in which the median-fit line intersects the horizontal axis?

Share & Summarize

    d. Describe how you determined the coordinates in part c. Your description should be written so that a classmate could use it to determine the coordinates. Be prepared to share your description with the class.

11. **a.** What would be the rental cost if Cedric rented a VCR and 12 tapes for a day?

    b. What would be the rental cost if Cedric rented a VCR and 35 tapes for a day?

    c. Would it make sense for him to rent 35 tapes? Explain.

    d. What would be the rental cost if Cedric rented a VCR and x tapes?

12. **a.** Write, in words, a rule that describes the relation between the number of movies rented, x, and the total cost of the rental, y, including one VCR.

    b. Compare your description with that of a classmate. Are they identical? similar? Do they say the same thing?

    c. Work together to produce a written description you both think is accurate and clear.

    d. Express your written description as an equation.

13. a. Use your rule to find Cedric's cost for renting one VCR and 15 tapes.

b. How much would it cost to rent the VCR and 26 tapes?

c. In addition to the VCR, how many tapes could Cedric rent if he had $25 to spend?

14. The speed limit on many U.S. highways is 65 miles per hour. How many miles would you travel if you drove at this speed limit for 1 hour? for 2 hours?

15. Copy and complete the table for miles traveled in hours at 65 miles per hour shown below.

Hours Driven (x)	1	2	3	4	5	6	7	8	9
Miles Traveled (y)	65								

16. a. Make a scatter plot for miles traveled versus hours driven. You may wish to use a software package to complete the scatter plot.

b. Draw the median-fit line for the scatter plot.

17. a. Use your graph to predict the miles traveled at 65 miles per hour if you drive 15 hours.

b. Are you confident in your prediction? Why?

18. Use your graph to find the miles traveled for the given hours of driving.

a. 10 hours

b. 2.5 hours

c. 1.75 hours

d. 3.25 hours

Share & Summarize

19. a. Write an equation for the relationship between the number of hours driven, x and number of miles traveled, y.

b. Use your equation to find the miles traveled in each part of Exercise 18. Be prepared to share your findings with the class.

● HOMEWORK PROJECT

20. Extension

    a. How could you use your graph to find the hours required to travel 300 miles at 65 miles per hour? What do you find?

    b. Use your equation to find the hours required to drive 400 miles at 65 mph. Describe the procedure you used.

21. Extension

    a. Cyndi, a salesperson at Betty's Pretty Good Boutique, works on commission. She makes $300 a week plus a commission of 5% of her total sales. How much does Cyndi earn in a week when she sells $4000 worth of clothing?

    b. Make a table showing Cyndi's weekly earnings for each additional $1000 in sales from $0 sales through $10,000 in sales.

    c. Use your table to make a scatter plot for weekly earnings versus total sales.

    d. Draw the median-fit line. Where does it intersect the axes?

22. Extension

    a. Write an equation for the relationship between Cyndi's weekly earnings if y is her earnings and x is her total weekly sales.

    b. What are Cyndi's least weekly earnings? When does it occur?

    c. What is Cyndi's weekly earnings if her sales are $2350?

    d. What were Cyndi's sales if she had earnings of $532.50?

23. Extension

    a. If Cyndi's salary were raised to $350 a week plus 5% commission on weekly sales, construct a graph of her weekly earnings for weekly sales from $0 to $10,000.

    b. How is this graph similar to the graph in Exercise 21? How is it different?

    c. Write an equation describing Cyndi's new weekly earnings. How is it similar to the one you wrote in Exercise 22 a.? How is it different?

Activity 4-2　Linear Equations

 graph paper

 ruler

 calculator

 software

● PARTNER PROJECT 1

1. a. John Clark sells appliances at Lowland Appliance Store. He earns $250 a week and a commission of 10% on all that he sells. Suppose y is his weekly earnings including the 10% commission on x dollars of sales. Find his earnings if x is $1500.

b. One week John's manager, Sheila Deppler, told John that he had sold $2500 in goods, but that Mr. Bumstead had returned $3000 in goods previously sold. What is John's sales total for the week?

c. What are John's earnings for the week described in Exercise 1b?

2. Copy and complete the table below of John's weekly sales and his earnings.

Sales: x	0	4000	5000	300	-500	2100	125	-4000	1300
Earnings: y	250	650							

3. a. Make a scatter plot for earnings versus sales on a sheet of graph paper. Choose your scale carefully. Label the axes and draw the median-fit line.

b. Describe the scale you chose for the grid. How did you label points to the left of 0 on the horizontal axis? How did you label points below 0 on the vertical axis?

● Share & Summarize

c. Compare your answer to part a and part b with that of your neighbor. Between the two of you, decide on a way to plot points where one or both of the coordinates are negative numbers. Be prepared to share your method with the class.

● GROUP PROJECT

4. a. Write a rule in the form of an equation that gives John's earnings, y, in terms of his base salary and his commission of 10% on sales of x dollars.

b. Use the equation to determine John's earnings for sales of $240, $850, $2340, −$440, and −$672.

c. Do the points calculated in part b lie on the median-fit line of Exercise 3a?

5. a. The equation you wrote in Exercise 4a is an example of a linear equation. Describe how this equation is used to calculate John's earnings.

   b. Why do you suppose the equation is called a "linear" equation? Be prepared to share your reasoning with the class.

6. a. The cost of operating a cellular telephone each month is given by a linear equation. If the base cost is $25 per month and $30 for each hour of use, write the linear equation giving the cost in terms of the number of hours of operation.

   b. Make a table that includes seven pairs of values for hours and cost per month for the use of a cellular telephone.

   c. Graph your seven pairs of values on graph paper.

   d. Do the points lie on a line? If so, draw it.

7. The perimeter *(P)* of a rectangle is the distance around the rectangle. For a rectangle with one dimension that is 10 inches long, the table gives the perimeter for several values of the other side, *s*.

Side: *s*	4	15	6	9	21	2	18	10
Perimeter: *P*	28	50	32	38	62	24	56	40

   a. Draw a rectangle and label the sides when $s = 4$ and $P = 28$.

   b. Find an equation relating the perimeter, *P*, to the length of the side, *s*.

   c. Graph the points on graph paper. Do they lie on a line? If so, draw it.

   d. At what point does the graph intersect the vertical axis?

   e. In your equation in part b, if *P* is 18, what is the value of *s*? Does this value for *s* make sense in this situation? Why?

8. a. The base, *b*, of a triangle is fixed at 10 units. The height of the triangle is represented by *h*. Let *A* represent the area of the triangle. Copy and complete the table of values relating the height and area. Remember that the area of a triangle is given by $A = \frac{1}{2}bh$.

Height: *h*	8	10	4	7	6	3	5	1.2	x
Area: *A*	40	50	20	35					

   b. Write an equation relating the area, *A*, and the height, *h*. Check your equation for several pairs in the table.

c. Use graph paper to graph the pairs in the table. Is the relation linear? If so, draw the line. Do all the points fall on this line?

d. Do negative values of h make sense in this situation? Explain.

9. a. The cost, C, of a gift is the price, p, plus the sales tax on the price. If the sales tax is 5%, make a table of the cost of gifts that are priced at $10, $15, $20, $25, $30, $40, and p.

b. Graph C versus p on graph paper.

c. Write an equation that relates C and p. Is it a linear equation?

10. a. The long distance cost for telephone calls at a certain time is given in the following table. Copy and complete the table.

Time (min):	1	2	3	4	5	6	7	x
Cost (cents):	25	43	61	79	97			

b. Write a rule for the cost of telephone service.

c. Graph the data on graph paper. Is the relation linear?

d. What is the cost of a 15-minute call? What is the cost of a 17-minute call?

11. Extension
a. How does the telephone company charge for long distance calls that are not an exact number of minutes?

b. Does your graph in Exercise 10 fit this method of charging?

Share & Summarize

c. Change your graph so that it gives the correct cost of a call whether it is a whole number of minutes or not. Be prepared to share your changes with the class.

● HOMEWORK PROJECT 1

Graphing Calculator Activity

You can learn how to use a graphing calculator to create tables in Activity 4 on page 61.

12. Extension
a. The expression $y = 10 - x$ defines a linear equation. Make a table of nine pairs of x and y values that satisfy the rule.

b. Graph the pairs. Do they lie on a line? If so, draw it.

13. Extension
a. The expression $y = x - 10$ defines a linear equation. Make a table of nine pairs of x and y values that satisfy the expression.

b. Graph the pairs. Do they lie on a line? If so, draw it.

14. Extension

a. Are the expressions in Exercises 12 and 13 the same? Explain their similarities and differences.

b. Are the graphs the same? Explain their similarities and differences.

● PARTNER PROJECT 2

15. Extension

a. For the data on male or female student athletes in Lindell High School on pages 5 and 6, are there any pairs of variables for one grade or for different grades whose points of the scatter plot should all lie on a line? Identify them. (Hint: Think about this. Do not try all sorts of combinations.)

b. Use graphing technology to draw the median-fit lines for the pairs of data you identified in part a. Do the lines confirm your conjecture? Explain.

16. Extension

a. Make a table of seven pairs of values of x and y that satisfy the expression $y = \dfrac{24}{x}$.

b. Do the points lie on a line? Explain your answer.

c. Would a line make a good summarizing shape for these data?

17. Extension

a. For the expression given in Exercise 16, is there a value of y that corresponds to x when it is 0? Explain.

b. Can y be equal to 0? Explain.

Portfolio Assessment

Select some of your work from this investigation that shows how you used a calculator or computer. Place it in your portfolio.

● HOMEWORK PROJECT 2

18. Extension

a. Describe the graphs of the linear equations $y = 5 + x$ and $y = x + 5$.

b. Describe the graphs of linear equations $y = x + c$ and $y = c + x$ where c is an arbitrary constant number.

19. Extension

Describe the graphs of the linear equations $y = x + 5$ and $y = x + 2$.

Journal

20. Journal Entry Write a linear equation of your choice relating x and y. Make a coordinate system and graph ten pairs of points that satisfy your equation. Find the median-fit line for these points. What can you say about the median-fit line and the points in the scatter plot?

Linear Functions

When the rule relating one variable to a second is a linear equation, the relation is a **linear function**. Engineers use linear functions to model physical situations. Business people model economic situations using linear functions. The distance you travel on your bicycle going a constant speed of 10 miles per hour can be represented by a linear function.

Activity 5-1 Equations, Tables, and Graphs

Materials

graph paper

calculator

software

● GROUP PROJECT

1. a. Sofia Lopez sells new cars. Her monthly salary is $1,500. She also is paid $50 for every car she sells. How many cars do you think she could sell in a good month? In a poor month?

   b. What is Sofia's monthly income for her "good month" sales? her "poor month" sales?

2. a. Sofia's monthly income, y, is a linear function of the number of cars, x, she sells. It is represented by the linear equation $y = 50x + 1,500$. What does the 1,500 represent? What does the 50 represent?

   b. Copy and complete the table below showing Sofia's monthly income as a function of the number of cars sold. How is the equation used to complete the table?

Number of Cars Sold: x	0	2	5	7	10	12	15	17	20
Sofia's Pay: y					2,000				

   c. Make a scatter plot and find the median-fit line.

d. Describe the relationship between the points and the line.

e. How many cars does Sofia need to sell to earn $4000?

**Share &
Summarize**

f. If Sofia sold n cars in a month, how much does she earn? Be prepared to explain your reasoning to the class.

3. For each linear function, make a table of four pairs of values and then graph the ordered pairs on graph paper. Do all the points lie on a line? Create one more pair satisfying the equation. Graph it and tell whether it lies on the same line.

   a. $y = 5x + 2$ **b.** $y = -3x + 4$ **c.** $y = 1.5x - 2$

● PARTNER PROJECT

Use a graphing calculator or computer function grapher to investigate Exercises 4-8.

4. a. Consider the equation $y = 2x + 1$. Copy and complete the table below by substituting the five different values of x into the right portion of the equation, and calculating the corresponding values of y. For example, if you let $x = -5$, then $2(-5) + 1$ is the value of y. In this case, $y = -9$. The first value in your table is shown.

**Graphing
Calculator
Activity**

You can learn how to use a graphing calculator to plot points in Activity 5 on page 62.

x	-5	-2	1	3	5	7
y	-9					

b. Use a graphing utility to plot the five points in the table above that satisfy the equation $y = 2x + 1$.

5. a. Examine the points plotted in Exercise 4. Do they appear to be on a line?

b. Now have your graphing calculator or computer display the graph of the equation $y = 2x + 1$ without erasing the points earlier graphed. What is this graph? How are the points and the line related?

**Graphing
Calculator
Activity**

You can learn how to use a graphing calculator to display a graph without erasing points graphed earlier and to trace a graph in Activity 6 on page 63.

6. a. Use the trace function of your graphing utility. Move the blinking cursor along the graph of $y = 2x + 1$ slowly. Do the coordinates of each of the points you graphed in Exercise 4 appear as coordinates of the trace point?

b. Move the cursor along to another position and write down the coordinates of the point. Do the coordinates satisfy the equation? Illustrate with examples.

Journal

7. a. Journal Entry Describe how you would use the graph of the equation $y = 2x + 1$ to find a value of y corresponding to a value of x.

b. Describe how you would use the equation to find a value of x corresponding to a known value of y.

8. For each equation, find the coordinates of three points that satisfy the equation, plot the three points using a graphing utility, and then draw the graph of the equation with the utility. Finally, indicate whether or not the points are on the graph.

a. $y = -2x + 1$ **b.** $y = 7x - 2$ **c.** $y = -4x + 3$

d. $y = 0.5x + 2$ **e.** $y = 1.5x - 3$ **f.** $y = x - 5$

9. a. Choose the correct term to complete the following sentence.

The graph of an equation like $y = 3x - 2$ is (always, sometimes, never) a straight line.

Share & Summarize

b. Be prepared to explain to the class why you made the choice you did in part a.

● **HOMEWORK PROJECT**

10. Extension

a. What do you suppose the graph of the equation $x = 2y - 1$ looks like?

b. How would you construct the graph? Describe your method and then use it to draw the graph. What is the result?

11. Extension

a. What do you suppose the graph of the equation $3x - 2y = 5$ looks like?

b. How would you construct the graph? Describe your method and then use it to draw the graph. What is the result?

Journal

12. Journal Entry   Investigate the graphs of equations of the form $Ax + By + C = 0$, where A, B, and C are real numbers and not both A and B are 0. Make a conjecture concerning the nature of the graphs, and then write an argument you would use with a friend to convince him or her that your conjecture is correct.

Activity 5-2 Visualizing Slopes and Intercepts

Materials

 graph paper

 colored pencils

 software

 Graphing Calculator Activity

You can learn how to use a graphing calculator to display a family of linear functions in Activity 7 on page 64.

 Share & Summarize

You may use whatever means available to you, such as graph paper and pencil, computer software, or graphing calculator, to explore these questions.

● PARTNER PROJECT

1. The graphs of functions like $y = 2x$ and $y = 7x$ are lines.

   a. Explore the family of linear functions with equations $y = mx$ where m is positive. Graph $y = mx$ using six different values between $\frac{1}{10}$ and 6 for m. Look for patterns in the resulting graphs. Display your graphs on the same coordinate axes.

   b. Explore the family of linear functions with equations $y = mx$ where m is negative using the same method as in part a.

   c. Make a conjecture about the role of m in the graph of $y = mx$. Compare your observations with those of a classmate. Be prepared to share your reasoning with the class.

2. Sketch a line and give an approximate value of m for a line containing the origin and lying:

   a. between $y = x$ and the x-axis   **b.** between $y = x$ and the y-axis

   c. between $y = -x$ and the x-axis   **d.** between $y = -x$ and the y-axis.

3. **a.** Explore the family of linear functions with equations $y = x + b$ by varying the value of b through both positive and negative values and looking for patterns in the resulting graphs. Display your graphs on the same coordinate axes.

   b. Explore the family of linear functions with equations $y = 4x + b$. Use several different values for b. Display your graphs on the same coordinate axes.

   c. Explore the family of linear functions with equations $y = -2x + b$. Use several different values for b. Display your graphs on the same coordinate axes.

   d. Make a conjecture about the role of b in the graph of $y = mx + b$. Compare your conjecture with that of a classmate.

Investigation 5 **Linear Functions**   **43**

4. Use your conjectures about m and b to describe the graph of each of the following equations. If possible, use a graphing utility to verify your descriptions.

a. $y = 5x - 6$

b. $y = -x + 0.4$

c. $y = 0.4x + 0.2$

d. $y = -10x - 8$

Share & Summarize

5. Journal Entry Summarize the results of the previous exercises. Be prepared to share your results with the class.

a. For a line with equation $y = mx + b$, what does m determine?

b. For a line with equation $y = mx + b$, what does b determine?

6. For a line with equation $y = mx + b$, m is the **slope**, and b is the **y-intercept.**

a. What is the slope and y-intercept of the graph of $y = 5x + 8$?

b. Write the equation of a different line that has slope 5.

c. Graph the equations in parts a and b on the same coordinate axes. How do the lines appear to be related?

d. Write the equation of a line different from that in part a that has a y–intercept of 8.

e. Graph the equations in parts a and d on the same coordinate axes. How are the lines related?

7. a. Write the equation of a line that has slope of 1 and intersects the two axes at the same point.

b. Draw the graph of your equation in part a.

c. Write an equation of a line that is steeper than that of the line in part a and that has a y-intercept of 0. Draw the graph of this equation on the same coordinate axes used for part b.

d. Write an equation of a line that is less steep than that of the line in part a and that has a y-intercept of 0. Draw the graph of this equation on the same coordinate axes used for part b.

⬤ GROUP PROJECT

8. Extension

a. What are the coordinates of the points where the graph of $3x - 2y = 6$ crosses the axes?

b. A **term** is a number, a variable, or a product or quotient of numbers and variables. Some examples of terms are 5, $\frac{ab}{4}$, and $7k$. The **coefficient** is the numerical part of a term. For example, in $7k$, the coefficient is 7. In $\frac{ab}{4}$ the coefficient is $\frac{1}{4}$. How are the coordinates in part a related to the coefficients in the equation?

9. Extension

a. Describe the graph of the equation $\frac{x}{3} + \frac{y}{4} = 1$.

b. At what points does the graph of $\frac{x}{3} + \frac{y}{4} = 1$ cross the x- and y-axes? How are the coordinates of these points related to the constants in the original equation?

10. Extension The graphs of the equations $y = 3x - 4$ and $y = -\frac{1}{3}x + 4$ are related in a special way. Investigate them and describe how they are related. How are the values of m in the equations related?

Activity 5-3 Calculating Slopes and Finding Equations

Materials

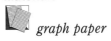
graph paper

Ski hills are often called "the slopes" because they have a large slope. A ski hill is shown in the figure below. The bottom of the hill has been given coordinates $(0, 0)$. Two other points on the line of the hill have been marked also. These are the points A$(400, 300)$ and B$(600, 450)$. Is this a steep hill, or is it gentle?

● PARTNER PROJECT

1. a. The table below contains ten points on the line with equation $y = 2x + 3$. Copy and complete the table.

x	−4		−2	−1	0		2	3	4	5
y	−5	−3	−1		3	5	7		11	

b. Compare the differences between pairs of *y*-coordinates with the differences between corresponding pairs of *x*-coordinates. What pattern do you see?

c. What is the slope of the line?

d. Is the slope related to the pattern you found in part b? How?

2. a. The table below contains eight points on the line with equation $y = 3x - 4$. Copy and complete this table.

x	−1	0	1		3	4		6
y	−7		−1	2	5		11	

b. Refer to the table in part a. Compare the differences between pairs of *y*-coordinates with the differences between pairs of corresponding *x*-coordinates. What pattern do you see?

c. What is the slope of the line?

d. Is the slope related to the pattern you found in part b? How?

3. a. The table below contains eight points on the line with equation $y = \frac{3}{2}x + 1$. Copy and complete this table.

x	−4	−2	0		4		8	10
y	−5	−2		4		10	13	

b. Compare the differences between pairs of *y*-coordinates with the differences between pairs of corresponding *x*-coordinates. What pattern do you see?

c. What is the slope of the line?

d. Is the slope related to the pattern you found in part b? How?

4. a. Combine the information you found in Exercises 1-3 to make a conjecture about a way to compute the slope of a line from two points on the line.

b. Compare your conjecture with a classmate. Work out any differences to develop a joint conjecture.

c. Use your conjecture to find the slope of the ski slope on page 45.

Share & Summarize

d. Is the slope steep or gentle? Be prepared to explain your reasoning to the class.

5. Find the slope of the line containing each pair of points below.

 a. $(3, 4)$ and $(5, 7)$ **b.** $(-1, 4)$ and $(5, 4)$ **c.** $(-4, -3)$ and $(-1, 2)$

 d. $(2, 5)$ and $(-2, -2)$ **e.** $(3, 3)$ and $(-2, -3)$ **f.** $(2, 4)$ and $(3, 2)$

6. For the lines given below, identify the coordinates of three points on the line. Calculate the slope of the line twice using different pairs of the three points. What result do you expect for the calculations for a given line?

 a. $y = 2x - 1$ **b.** $y = -3x - 1$ **c.** $y = -4x - 9$

Journal

7. Journal Entry Would you agree or disagree with this statement: "If I choose two points on a line and calculate the slope and you do the same with two other points on the line, the values for the slopes we get will be identical." Explain your position.

8. a. If a line has a negative slope, does it fall to the right or rise to the right on a coordinate system?

b. What kind of line has a slope of 0?

GROUP PROJECT

Since two points determine a line and a line has a unique slope given by m in the equation $y = mx + b$, you can write the equation of a line whenever you know two points.

9. a. Suppose you know that the points $(1, 1)$ and $(4, 7)$ lie on a line. Devise a way to write the equation of the line containing these points in the form $y = mx + b$. (Remember that m is the slope of the line and that b is the y-intercept of the line.)

b. Summarize the procedure you developed. Compare your procedure with that of another group.

c. Are the procedures the same? Are they similar? Do they both work?

d. Which procedure is easiest to do? Which method will you use? Be prepared to share your procedure with the class.

10. Use the procedure you developed in Exercise 9 to find the equation of the line containing each pair of points.

   a. (4, 4) and (5, 7)   **b.** (4, 2) and (5, 4)   **c.** (−4, −3) and (−1, 3)

   d. (2, 6) and (−2, −2)   **e.** (3, 3) and (−2, −3)   **f.** (2, 4) and (3, 2)

11. Extension The y-intercept of a line is 4, and the point (−2, 2) is on the line. Find the equation of the line.

12. Extension The y-intercept of a line is −2, and the x-intercept is 4. Find the equation of the line.

13. Extension The slope of a line is −2, and the point (3, 5) is on the line. Find the equation of the line.

14. Extension You read in the paper that a new highway will have a grade of 3%. What will be the slope of that road?

15. Extension Make a general argument, that, if a line has equation $y = mx + b$, then the slope of that line is m.

HOMEWORK PROJECT

16. On a sheet of graph paper, sketch a line for each situation below. In each case, explain why there is only one such line or why there are many lines with the given characteristic(s).

**Portfolio
Assessment**

Select an item from your work that shows your creativity and place it in your portfolio.

   a. The line has a slope greater than 1.

   b. The line has a y-intercept of 3.

   c. The line has a slope between 0 and 1 and a y-intercept of 2.

   d. The line has a slope less than −1 and a y-intercept of 0.

   e. The line contains the points (1, −2) and (3, 4).

   f. The line has a slope between 0 and −1.

   g. The line has a slope of −3 and a y-intercept of 2.

Using Linear Functions

Median-fit lines are useful in exploring paired data such as that of Athletic Director Molar's student data. A helpful final technique is to write an equation of the line used to summarize those data. With this equation, you can substitute an initial value and get the predicted value immediately.

Activity 6-1   Equations for Fitted Lines

Materials

graph paper

calculator

software

tape measure

● GROUP PROJECT

1. Select 12 or 13 classmates and measure their height and stride lengths. Record your measurements in a table like the one shown below. You may need to convert inches to centimeters by multiplying by 2.54.

Name	Height (cm)	Stride Length (cm)

2. a. Make a scatter plot of stride length versus height.

b. Is the pattern of the data linear enough to be summarized by a line? Are there any outliers (made by classmates who did not walk naturally)? If so, reject those points, and construct the median-fit line for the graph.

3. a. Determine the equation of the median-fit line by determining the slope and the *y*-intercept.

b. Is the slope positive or negative? What is the slope?

c. What does the slope represent in terms of the data?

d. Is the *y*-intercept positive or negative? Estimate the *y*-intercept from the graph.

e. What does the *y*-intercept represent in terms of the data? Does it make sense?

4. Use the graph of your line to predict the length of stride for a person with the given height.

    a. 150 cm   **b.** 160 cm   **c.** 170 cm   **d.** 180 cm

5. a. Use statistics software to produce a scatter plot and the median-fit line.

    b. Compare the equation of the line produced by the software with your equation. How do they differ? Are the slopes similar? Are the y-intercepts close?

6. a. Use the data for the entire class and the statistics software to produce a scatter plot and summarizing line. What is the equation of this line?

Share & Summarize

    b. Use this line to predict the stride length associated with each of the four heights in Exercise 4. Be prepared to share your findings with the class.

7. a. Use the data in the three tables below to make scatter plots for Winning time versus Year.

Women's 400-meter Freestyle Relay, 1912-1988

Year	Country	Time
1912	Great Britain	5:52.80
1920	United States	5:11.60
1924	United States	4:58.60
1928	United States	4:47.60
1932	United States	4:38.00
1936	Holland	4:36.00
1948	United States	4:29.20
1952	Hungary	4:24.40
1956	Australia	4:17.10
1960	United States	4:08.90
1964	United States	4:03.80
1968	United States	4:02.50
1972	United States	3:55.19
1976	United States	3:44.82
1980	East Germany	3:42.71
1984	United States	3:43.43
1988	East Germany	3:40.63

Source: Information Please Almanac, 1994

Women's 100-meter Butterfly, 1956-1988

Winner	Time
1956: Shelly Mann, United States	1:11.00
1960: Carolyn Schuler, United States	1:09.50
1964: Sharon Stouder, United States	1:04.70
1968: Lynn McClements, Australia	1:05.50
1972: Mayumi Aoki, Japan	1:03.34
1976: Kornelia Ender, East Germany	1:00.13
1980: Caren Metschuck, East Germany	1:00.42
1984: Mary T. Meagher, United States	59.26
1988: Kristin Otto, East Germany	59.00

Source: Information Please Almanac, 1994

Men's 400-meter Relay, 1912-1988 (Track)

Year	Country	Time (sec)
1912	Great Britain	42.40
1920	United States	42.20
1924	United States	41.00
1928	United States	41.00
1932	United States	40.00
1936	United States	39.80
1948	United States	40.60
1952	United States	40.10
1956	Germany	39.50
1960	United States	39.50
1964	United States	39.00
1968	United States	38.20
1972	United States	38.19
1976	United States	38.33
1980	Soviet Union	38.26
1984	United States	37.83
1988	Soviet Union	38.19

Source: Information Please Almanac, 1994

b. Find the equation of the summarizing line for each scatter plot.

c. Use the equations of part b to predict the 1992 winning time for each event.

d. Compare your predictions with the actual 1992 winning times.

8. a. What is the slope of each summarizing line in Exercise 7?

b. Describe what the slope tells you about winning times as the years change.

c. Where does the summarizing line cross the *x*-axis? What does this point represent in terms of winning time and year? Does this make sense? Explain.

⬤ PARTNER PROJECT

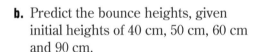

9. a. Get two balls designed for different sports such as a tennis ball, a basketball, or a golf ball. Choose ten different heights from which to drop the balls. For each drop, measure the height of the first and second bounces. Record your data in a table and construct a scatter plot for the height of the first bounce versus the original height. Next construct a scatter plot of height of second bounce versus the original height. Determine a summarizing line and its equation for each scatter plot. Compare the two equations.

b. Predict the bounce heights, given initial heights of 40 cm, 50 cm, 60 cm and 90 cm.

c. Can you predict the height of the second bounce from the scatter plot for first bounce versus original height? Explain.

d. Compare the lines generated for the two different balls. Are they similar? Explain.

Graphing Calculator Activity

You can learn how to use a graphing calculator to find equations of summarizing lines in Activity 8 on page 65.

10. Extension

a. Find an equation of a summarizing line for twelfth grade 40-yard dash time versus ninth grade 40-yard dash time for males or females from the data on pages 5 and 6.

b. If each twelfth grade time was decreased by 0.1 second, what would be the equation of the summarizing line for these data? How is it related to the equation of the original line?

c. Increase each time in the twelfth grade by 0.2 second, and find the equation of the summarizing line. How is it related to the equations of the other lines?

d. Make a conjecture about the effects of adding or subtracting a constant to or from each *y*-coordinate value in relation to the equation of the summarizing line.

11. Extension

    a. Using the same data you used in Exercise 10a, add 0.1 second to each ninth-grade time, and find the equation of the summarizing line for the resulting scatter plot.

    b. Now subtract 0.2 second from each ninth-grade value and find the equation of the summarizing line for the resulting scatter plot.

    c. Describe the effect on the summarizing line when the values of each *x*-coordinate are changed by adding a constant.

12. Extension

    a. Given an equation $y = mx + b$, how is the graph of $y + c = mx + b$ related to the graph of $y = mx + b$? Verify your conjecture.

    b. How is the graph of $y - c = mx + b$ related to the graph of $y = mx + b$? Verify your conjecture.

13. Extension

    a. Given an equation $y = mx + b$, how is the graph of $y = m(x + c) + b$ related to the graph of $y = mx + b$? Verify your conjecture.

    b. How is the graph of $y = m(x - c) + b$ related to the graph of $y = mx + b$? Justify your conjecture.

Activity 6-2   Manipulating Linear Equations

Materials

cups and counters

software

    Often when working with linear equations you know a value for *y* and want to find a value for *x* . A procedure for dealing with this kind of situation is to model equations using cups and counters. In this model, a cup represents the variable, yellow counters ($\oplus$) represent positive integers, and red counters ($\ominus$) represent negative integers.

A model for the expression $3x + 2$ is shown at the right.

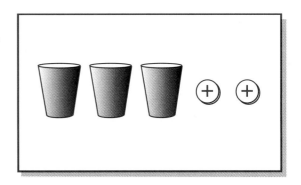

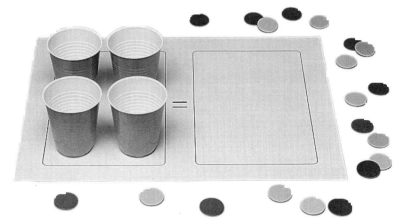

● PARTNER PROJECT

1. Model each expression with cups and counters.

   a. $2x + 5$ **b.** $4x + 7$

   c. To model $2x - 3$, first rewrite $2x - 3$ as $2x + (-3)$. Then $2x + (-3)$ can be modeled as follows.

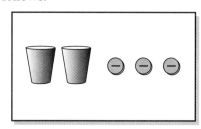

2. Model $3x - 5$ with cups and counters.

3. What expression is modeled by each model below?

   a.
   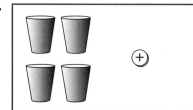

   b.

   c.

   d.

4. Cups and counters can also be used to model equations. The equation $2x + 5 = 9$ is modeled below.

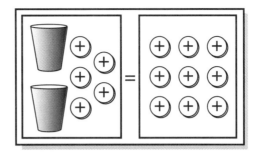

a. If you replaced each cup with a positive counter, would the left side still equal the right side? Explain.

b. If you replaced each cup with a negative counter, would the left side still equal the right side? Explain.

c. Think of the cups as being weights on each side of a balance scale. How many counters would balance one cup? That is, determine the value of x for which this statement is true. Check your result by substituting for x in the original equation.

d. Describe the steps in your procedure.

e. Use cups and counters to solve $3x + 4 = 13$.

f. Solve $5x + 3 = 18$ for x without using cups and counters. What was your first step? your second step?

5. Cups and counters can also be used to model equations like $2x - 5 = 9$. First write $2x - 5 = 9$ as $2x + (-5) = 9$. Then model the equation as follows.

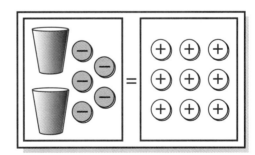

a. Use the model to help you solve the equation.

b. Solve the equation $4x - 2 = 10$ using cups and counters.

c. Describe the steps you used in part b. Can you use your procedure to solve an equation without using cups and counters? Try it on $3x - 7 = 3$.

6. a. Compare the steps in your procedure for solving an equation like $3x - 5 = 23$ with those of others in your class. Are they the same?

b. If the procedures are different, which way is easier for you to use? Which way is easier for you to understand? Do they both give the same result for x? If not, can they both be correct?

Share & Summarize

c. Journal Entry Choose the procedure you understand most clearly for use in solving equations. Give a written description of its steps. Be prepared to share your reasons with the class.

● HOMEWORK PROJECT

7. Copy each table below. Apply the procedure you chose in Exercise 6c to determine x for each value of y in the following equations.

a. $y = 3x - 3$ **b.** $y = -2x + 1$ **c.** $y = \frac{2}{3}x - 1$

x	y
	3
	-2
	0
	0.2
	-0.4
	$\frac{3}{4}$

x	y
	3
	-2
	0
	0.2
	-0.4
	$\frac{3}{4}$

x	y
	3
	-2
	0
	0.2
	-0.4
	$\frac{3}{4}$

8. a. Using the Lindell High School data on ninth- and twelfth-grade times in the 40-yard dash for males or females, found on pages 5 and 6, construct a scatter plot of twelfth-grade times versus ninth-grade times.

b. Determine the equation of the summarizing line (median-fit).

9. Use your equation from Exercise 8b to predict ninth-grade times given the following twelfth-grade times.

a. 5.1 s

b. 5.8 s

c. 4.9 s

d. 6.1 s

10. Use your equation from Exercise 8b to predict twelfth-grade times when given the following ninth-grade times.

a. 6.5 s

b. 6.0 s

c. 5.8 s

d. 5.3 s

GROUP PROJECT

11. a. Use the data on pages 5 and 6 to construct a scatter plot for twelfth grade bench-press (or leg-press) weight versus ninth grade bench-press (or leg-press) weight.

Portfolio Assessment

Review items in your portfolio. Make a table of contents of the items, noting why each item was chosen. Replace any items that are no longer appropriate.

b. Determine the equation of the summarizing line.

12. Use your equation from 11b to predict ninth-grade values for the following twelfth-grade values.

a. 180 lb

b. 210 lb

c. 165 lb

d. 250 lb

13. Use your equation from Exercise 11b to predict twelfth-grade values for the following ninth-grade values.

a. 120 lb

b. 145 lb

c. 160 lb

d. 100 lb

Journal

14. Journal Entry Describe how an equation of a summarizing line for a scatter plot, in the form $y = mx + b$, can be used:

a. to find y when x is given.

b. to find x when y is given.

Graphing Calculator Activities

Graphing Calculator Activity 1: Scatter Plots

One way to determine whether there is a relationship between two sets of data is to display the information in a scatter plot. Graphing calculators are capable of drawing scatter plots for data that you enter into the memory.

Example The elevation and number of clear days in 1992 for several U.S. cities are shown in the table below. Use a TI-82 graphing calculator to create a scatter plot of the data.

City	Elev. (ft)	Clear days	City	Elev. (ft)	Clear days	City	Elev. (ft)	Clear days
Albany, NY	275	67	Dallas, TX	551	118	New Orleans, LA	4	83
Atlanta, GA	1010	110	Denver, CO	5283	113	Pittsburgh, PA	1137	55
Bismarck, ND	1647	90	Fresno, CA	328	170	Seattle, WA	400	67
Boston, MA	15	90	Nashville, TN	590	88	Springfield, MO	1268	103

Before you create a scatter plot, you must clear the statistical memories.

Enter: [STAT] 4 [2nd] [L1] [ENTER] [STAT] 4 [2nd] [L2] [ENTER]

Next, enter the data. Enter the elevations in list L1 and the clear days in list L2.

Enter: [STAT] [ENTER] *Accesses the statistical lists.*

275 [ENTER] 1010 [ENTER] ... 1268 [ENTER]

[▶] 67 [ENTER] 110 [ENTER] ... 103 [ENTER]

After the data is entered, the range for the graph must be set. A viewing window of [0, 5500] by [0, 200] with a scale factor of 500 on the x-axis and 25 on the y-axis is appropriate for this data.

Enter: [WINDOW] [ENTER] 0 [ENTER] 5500 [ENTER] 500 [ENTER] 0

[ENTER] 200 [ENTER] 25 [ENTER]

Now, you may choose the type of statistical graph and create the graph.

Enter: [2nd] [STAT PLOT]

● Try This

The table below shows the average heights in feet of 30 men and 30 women at different ages. Use a TI-82 graphing calculator to create a scatterplot for the data.

Age (yr)	1	3	5	10	12	15	18	20	22
Men	2.4	3.2	3.8	4.5	4.8	5.3	5.7	5.9	6.0
Women	2.5	3.3	3.7	4.4	4.9	5.2	5.3	5.4	5.5

Graphing Calculator Activity 2:   The Correlation Coefficient

When you are analyzing a set of data, it is often difficult to determine a relationship with a quick comparison. Finding the correlation coefficient of the data can be very helpful. You can use a graphing calculator to find the correlation coefficient of a set of data you enter.

Example The table below shows the rank in area and the rank in order of when each of the contiguous United States entered the Union. Find the correlation coefficient of the data.

State	Area	Order	State	Area	Order	State	Area	Order	State	Area	Order
AL	28	22	IA	24	29	NE	14	37	RI	48	13
AZ	5	48	KS	13	34	NV	6	36	SC	39	8
AR	26	25	KY	36	15	NH	43	9	SD	15	40
CA	2	31	LA	30	18	NJ	45	3	TN	33	16
CO	7	38	ME	38	23	NM	4	47	TX	1	28
CT	46	5	MD	41	7	NY	29	11	UT	10	45
DE	47	1	MA	44	6	NC	27	12	VT	42	14
FL	21	27	MI	22	26	ND	16	39	VA	35	10
GA	20	4	MN	11	32	OH	34	17	WA	19	42
ID	12	43	MS	31	20	OK	17	46	WV	40	35
IL	23	21	MO	18	24	OR	9	33	WI	25	30
IN	37	19	MT	3	41	PA	32	2	WY	8	44

Clear lists L1 and L2 in the statistical memory before you enter the data.

Enter: STAT 4 2nd L1 ENTER STAT 4 2nd L2 ENTER

Now enter the data. Enter the area in list L1 and the order in list L2.

Enter: STAT ENTER *Accesses the statistical lists.*

28 ENTER 5 ENTER 26 ENTER ... 8 ENTER

▶ 22 ENTER 48 ENTER 25 ENTER ... 44 ENTER

The TI-82 graphing calculator is capable of finding many different correlation values. The correlation coefficient we will use is the Pearson-product moment correlation.

Enter: STAT ▶ 9 ENTER

The correlation coefficient for the area and order of the contiguous United States is -0.8030829353. The data has a very strong negative relationship.

● Try This

Find the correlation coefficient to the nearest hundredth.

	Millions of Elementary and Secondary School Students									
Year	1900	1910	1920	1930	1940	1950	1960	1970	1980	1990
Students	10.6	12.6	16.2	21.3	22.0	22.3	32.5	42.5	38.2	38.0

Graphing Calculator Activity 3:   Median-Fit-Lines

When data is collected in real-life situations, the relationship between two sets of values is rarely a straight line. However, the relationship may be approximated by a straight line. One way to approximate the graph of data is by finding the median-fit line. Your graphing calculator can find the median-fit line for data you enter.

Example   Draw a scatter plot and the median-fit line for the data about the orbits of ten asteroids that is given in the table below.

Asteroid	Ceres	Pallas	Juno	Vesta	Astraea	Hebe	Iris	Flora	Metis	Hygeia
Mean Distance from Sun (millions of miles)	257.0	257.4	247.8	219.3	239.3	225.2	221.4	204.4	221.7	222.6
Orbital period (years)	4.60	4.61	4.36	3.63	4.14	3.78	3.68	3.27	3.69	5.59

Begin by clearing the statistical lists and entering the data.
Enter: [STAT] 4 [2nd] [L1] [ENTER] [STAT] 4 [2nd] [L2] [ENTER] [STAT] [ENTER] 257.0 [ENTER] 257.4 [ENTER] ... 222.6 [ENTER] [▶] 4.60 [ENTER] 4.61 [ENTER] ... 5.59 [ENTER]

Enter the viewing window for the scatter plot. The data suggests a window of [200, 260] by [3, 6] with a scale factor of 10 for the x-axis and 0.5 for the y-axis.

Enter: [WINDOW] [ENTER] 200 [ENTER] 260 [ENTER] 10 [ENTER] 3 [ENTER] 6 [ENTER] 0.5 [ENTER]

Create the scatter plot by pressing [2nd] [STAT PLOT] [ENTER] and then using the arrow and [ENTER] keys to highlight "On", the scatter plot, L1 as the Xlist, L2 as the Ylist, and • as the mark.

Use the [CLEAR] key to clear any equations that are already in the Y= list before you enter the equation of the median-fit line.

Find the equation of the median-fit line by pressing [STAT] [▶] 4 [ENTER] . Press [Y=] [VARS] 5 [▶] [▶] 7 to add the equation to the Y= list. Press [GRAPH] to see the scatter plot and the median-fit line.

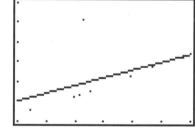

● Try This

The table below shows the percent of American television owners who had cable television in different years. Create a scatterplot and graph the median-fit line for the data.

Year	Percent	Year	Percent	Year	Percent	Year	Percent
1977	16.1	1981	28.3	1985	46.2	1989	57.1
1978	17.9	1982	35.0	1986	48.1	1990	59.0
1979	19.4	1983	40.5	1987	50.5	1991	60.6
1980	22.6	1984	43.7	1988	53.8	1992	61.5

Graphing Calculator Activity 4:   Tables

A graphing calculator is a powerful tool for studying functions. One way you can examine a function is to create a table of values. A TI-82 graphing calculator will allow you to create a large table of values quickly.

Example As a thunderstorm approaches, you see lightening as it occurs, but you hear the accompanying sound of thunder a short time afterward. The distance y in miles, that sound travels in x seconds is given by the equation $y = 0.21x$. Use a graphing calculator to create a table of values for x = {0, 0.5, 1, 1.5, 2, 2.5, 3, 3.5, ...}. How far away is lightening when the thunder is heard 3 seconds after the light is seen?

First, enter the function $y = 0.21x$ into the Y= list. Press $\boxed{Y=}$ to access the list. Then enter the equation in as function Y1. Use the $\boxed{CLEAR}$ key to remove any equations that are already in the list.

Now press $\boxed{2nd}$ $\boxed{TblSet}$ to display the table setup menu. The table is to start at 0, so enter 0 as the TblMin value and press $\boxed{ENTER}$. ΔTbl is the change between each pair of successive x-values in the table. Enter 0.5 as the ΔTbl value and press $\boxed{ENTER}$. Use the arrow and keys to highlight "Auto" for both the dependent and independent variables so that the calculator will construct the table automatically. Press $\boxed{2nd}$ $\boxed{Table}$ to display the completed table.

Use the arrow keys to scroll through the table entries. According to the table, when thunder is heard 3 seconds after the lightening is seen, the lightening is 0.63 miles away.

X	Y$_1$	
0	0	
.5	.105	
1	.21	
1.5	.315	
2.	.42	
2.5	.525	
3	.63	

X = 3

● Try This

1. Geothermal energy is generated whenever water comes in contact with heated underground rocks. The heat turns the water into steam that can be used to make electricity. The underground temperature of rocks varies with their depth below the surface. The temperature, y, in degrees Celsius is estimated by $y = 35x + 20$, where x is the depth in kilometers.

a. Use a graphing calculator to create a table of values for x = {0, 5, 10, 15, 20, 25, 30, 35, ...}.

b. What would be the temperature of rocks that are 15 kilometers deep?

2. The distance, y, in feet that an object falls in x seconds is found by $y = 16x^2$.

a. Use a graphing calculator to create a table of values for x = {0, 0.5, 1, 1.5, 2, 2.5, 3, 3.5, ...}.

b. How far will an object fall in 7 seconds?

Graphing Calculator Activity 5:   Plotting Points

The graphics screen of a graphing calculator can represent a coordinate plane. The *x*- and *y*-axes are shown, and each point on the screen is named by an ordered pair. You can plot points on a graphing calculator just as you do on a coordinate grid.

The program below will plot points on the graphics screen. In order to use the program, you must first enter the program into the calculator's memory. To access the program memory, use the following keystrokes.

Enter: [PRGM] [▶] [▶] [ENTER]

Example   Plot the points (0, −4), (3, 1), (−5, −4), (−1, 6), and (8, 2) on a graphing calculator.

First, set the range. The notation [−10, 10] by [−8, 8] means a viewing window in which the values along the *x*-axis go from −10 to 10 and the values along the *y*-axis go from −8 to 8.

Enter: [WINDOW] [ENTER]

[(−)] 10 [ENTER] 10 [ENTER] 1 [ENTER]

[(−)] 8 [ENTER] 8 [ENTER] 1 [ENTER]

```
Prgm1: PLOTPTS
:FnOff
:PlotsOff
:ClrDraw
:Lbl 1
:Disp "X="
:Input X
:Disp "Y="
:Input Y
:Pt-On(X, Y)
:Pause
:Disp "PRESS Q TO QUIT,"
:Disp "1 TO PLOT MORE"
:Input A
:If A = 1
:Goto 1
```

The program is written for use on a TI-82 graphing calculator. If you have a different type of programmable calculator, consult your User's Guide to adapt the program for use on your calculator.

Now run the program.

Enter: [PRGM] 1 [ENTER]

Enter the coordinates of each point. They will be graphed as you go. Press [ENTER] after each point is displayed to continue in the program.

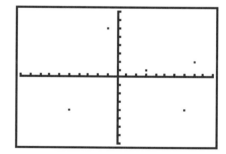

● Try This

Use the program to graph each set of points on a graphing calculator. Then sketch the graph.

1. (7, −1), (−3, 6), (−1, 2), (6, 8)

2. (−1.7, 2.2), (0.8, 1.9), (−1.2, 0.1), (−2.1, −3.7), (1.6, 3.2)

3. (−32, 4), (25, −15), (−13, −18), (−5, −11)

4. (92, 40), (−67, 21), (−51, −37), (24, −16), (32, −57), (89, 21)

Graphing Calculator Activity 6:   Tracing a Function

You can explore the characteristics of different functions by observing their graphs. The first step to graphing a function on a graphing calculator is to set an appropriate range. A viewing window of $[-10, 10]$ by $[-10, 10]$ with a scale factor of 1 on both axes denotes the domain values $-10 \le x \le -10$ and the range values $-10 \le y \le 10$. The tick marks on both axes in this viewing window will be one unit apart. This is called the standard viewing window. The standard viewing window is a good place to start when graphing an unfamiliar function.

Example   Graph $y = 2x + 3.7$ in the standard viewing window. Then use the trace function to determine whether the point $(-1, 2.3)$ is on the graph.

First, enter the function into the Y= list. If any functions are already on the list, clear them by using the arrow keys to move the cursor anywhere in the equation and then pressing the CLEAR key.

Enter: Y= 2 X,T,θ + 3.7

Now, select the standard viewing window and graph. The TI-82 graphing calculator has the standard viewing window as a choice on the zoom menu, so you can choose it without entering the range values manually.

Enter: ZOOM 6   *Selects the standard viewing window and completes the graph.*

You can use the trace function of a graphing calculator to find approximations of the coordinates of points that appear on the graph of a function. Press TRACE to get a blinking cursor on the graph. The right and left arrow keys allow you to move the cursor along the graph. Move the cursor to a point on the line where the *x*-coordinate is as close as possible to -1. Since the *y*-coordinate is not close to 2.3, we can determine that $(-1, 2.3)$ is not on the graph of $y = 2x + 3.7$.

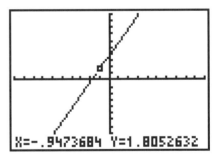

X=-.9473684 Y=1.8052632

● Try This

Graph each function in the standard viewing window. Then determine whether the given point is on the graph of the function.

1. $y = 3x - 2.1$; $(6, 6.2)$ **2.** $y = -2x - 4$ $(-3, 2)$
3. $y = 6 - x$; $(7, -1)$ **4.** $y = 2.5 + 3x$; $(-3, -6.5)$
5. $y = 11x - 6.9$; $(1, 7.4)$ **6.** $y = 13 - 5.5x$; $(2.5, 1.5)$
7. $y = x + 5.93$; $(4, 9.93)$ **8.** $y = -0.5x + 4.22$; $(8.44, 0)$

Graphing Calculator Activity 7: Families of Linear Equations

A family of graphs is a group of graphs that displays one or more similar characteristics. Many linear functions are related because they have the same slope or the same y-intercept as other functions in the family. All linear functions can be written in the form $y = mx + b$, where m represents the slope of the line and b is the y-intercept. You can graph several functions on the same screen and observe if any family traits exist.

Example   Graph the following functions on the same screen in the standard viewing window. Then describe the family of graphs to which they belong.

$y = 0.5x + 1$ $y = 3x + 1$

$y = x + 1$ $y = 5x + 1$

The TI-82 graphing calculator allows you to graph up to nine functions at one time. Enter each equation into the Y= list. *Be sure to clear any equations that are in the list before you start.*

Enter: [Y=] 0.5 [X,T,θ] [+] 1 [ENTER] [X,T,θ] [+] 1 [ENTER]

3 [X,T,θ] [+] 1 [ENTER] 5 [X,T,θ] [+] 1

The standard viewing window can be selected automatically from the zoom menu. Press [ZOOM] 6. The graphs will appear automatically.

When all the functions are graphed on the same screen, you can observe that they are all lines and that they all pass through the point (0, 1). However, their slopes are different. If you compare each graph with its slope, you find the greater the slope, the greater the angle formed by the line and the x-axis. This family of graphs is described as lines that have a y-intercept of 1.

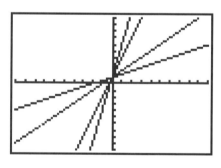

● Try This

Graph the following functions on the same screen. Then describe the family of graphs to which they belong.

1. $y = 3x$ **2.** $y = 4$ **3.** $y = 2x - 4$

$y = 3x + 1$ $y = -1$ $y = 4x - 4$

$y = 3x - 2$ $y = 2$ $y = -3x - 4$

$y = 3x + 5$ $y = 5.2$ $y = -x - 4$

Graphing Calculator Activity 8: Finding the Equation of a Median-Fit Line

Graphing calculators are capable of finding the median-fit line for a set of data that you enter into the statistical memory. Once the equation of the median-fit line is found, you can use the equation to approximate solutions to other problems.

Example The table below shows median age of the resident population of the United States for different years. Find the equation of the median-fit line of the data. Then use the equation to predict the median age in the year 2010.

Year	Median Age	Year	Median Age	Year	Median Age
1820	16.7	1900	22.9	1970	28.0
1840	17.8	1920	25.3	1980	30.0
1860	19.4	1930	26.4	1983	30.8
1870	20.2	1940	29.0	1984	31.1
1880	20.9	1950	30.2	1985	31.4
1890	22.0	1960	29.5	1990	32.9

First, clear the statistical lists and enter the values. Use list L1 for the years and list L2 for the populations.

Enter: [STAT] 4 [2nd] [L1] [ENTER] [STAT] 4 [2nd] [L2] [ENTER]
[STAT] [ENTER] 1820 [ENTER] 1840 [ENTER] ... 1900 [ENTER]
[▶] 16.7 [ENTER] 17.8 [ENTER] ... 32.9

The TI-82 graphing calculator will find many different statistical equations. The median-fit line is one of the choices available on the statistical calculation menu.

Enter: [STAT] [▶] 4 [ENTER]

The calculator will display the values of a and b for a median-fit line of the form $y = ax + b$. To the nearest thousandth, the values for this data are $a = 0.094$ and $b = -155.245$. Thus, the equation of the median-fit line is $y = 0.094x - 155.245$. Substituting 2010 for x in the equation, gives an estimate of 33.695 for the median age in the year 2010.

● Try This

Find the equation of the median-fit line. Then predict the percentage of calories from carbohydrates for a food with 27% calories from fat.

Food	% Calories from Fat	% Calories from Carbohydrates
Apple (medium)	9	89
Bagel (plain)	6	76
Banana (medium)	2	93
Bran Muffin (large)	40	53
Fig Newtons (4 bars)	18	75
PowerBar (any flavor)	8	76
Snickers bar (regular size)	42	51
Ultra Slim-Fast bar (one)	30	63

Glossary

A

Associated (p. 8)   Two sets of data are associated if an increase in one is accompanied by an increase in the other or an increase in one is accompanied by a decrease in the other.

Example   Two variables such as height and shoe size are associated because an increase in one variable accompanies an increase in the other.

C

Coefficient (p. 45)   The numerical part of a term.

Example   In 8*ab* the coefficient is 8.

Correlation (p. 13)   An interdependence between sets of data.

Example   The data represented in a scatter plot whose points either rise or fall from left to right have a correlation.

Correlation (p. 14)   A number that indicates the degree of correlation between two coefficient sets of data.

Example   A correlation coefficient of 0.9 represents a strong positive association, while a - 0.01 represents a weak negative association.

F

Fitting a line to the data (p. 16)   Drawing a line to summarize a scatter plot so that there are about as many points above the line as below it and that the line contains some of the points.

Example

Data Insights: Scatter Plot

MEN'S OLYMPIC 100-METER DASH TIMES

$Y = -0.01176X + 33.23726$

L

Linear function (p. 40)   A relation whose rule relating one variable to a second is a linear equation.

Example   The equation $y = 2x + 1$ represents a linear function.

M

Mathematical model (p. 17)   A representation of data that can be used to predict one value of a pair of data when given the other value.

Example   A visually-fit line is an example of a mathematical model for a pair of data.

Mean residual (p. 22)   A measure of how well a summarizing line fits the data. It is the sum of the residuals of a summarizing line divided by the number of residuals of the line.

Example   If the sum of the 12 residuals of a line is 6, the mean residual is $6 \div 12$ or 0.5. A relatively small mean residual such as this indicates that a summarizing line is a good fit to the data.

Median-fit line (p. 24)   A summarizing line that is drawn using three medians in a set of data.

Example

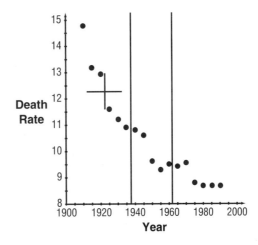

Median point (p. 26)   A point whose *x*-coordinate is the median of the *x*-coordinates of all the points in a set of data and whose *y*-coordinate is the median of the *y*-coordinates of all the points in a set of data.

Example   A set of data that contains the points (2, 3), (1, 0), (3, 4) has a median point of (2, 3).

N

Negative association
(p. 9)

The pattern on a scatter plot where the points get closer to the horizontal axis as the values on the horizontal axis increase.

Example

A scatter plot of *the value of a car versus the age of a car* will have a negative association.

No association (p. 9)

The pattern on a scatter plot where the points are scattered all over the graph and a reliable prediction cannot be made.

Example

A scatter plot of *height versus birth month* will have no association.

Perfect association
(p. 13)

A pattern on a scatter plot where all the points fall on a line.

Example

A scatter plot of *number of hours worked versus gross pay* will have perfect association.

Positive association
(p. 9)

The pattern on a scatter plot where the points get further away from the horizontal axis as the values on the horizontal axis increase.

Example

A scatter plot of *age versus height* will have positive association.

R

Residual (p. 22)

The vertical distance from a point to a summarizing line in a scatter plot.

Example

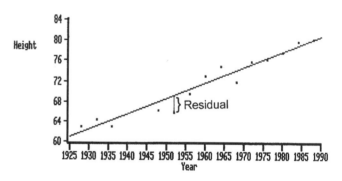

S

Scatter plot (p. 7)   A graph on a coordinate system that shows the relationship between two sets of data.

Example   The graph below is a scatter plot of *40-yd dash time versus body weight.*

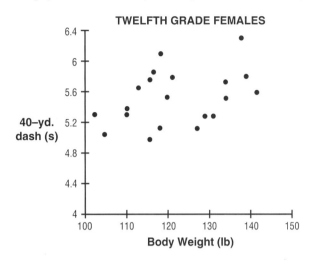

Slope (p. 44)   The slope of a line is the difference between the *y*-coordinates of any two points on the line divided by the difference between the corresponding *x*-coordinates. The slope of a line described by $y = mx + b$ is m.

Example   The slope of the equation of the line $y = 3x - 2$ is 3.

Spearman rank correlation (p. 15)   The correlation coefficient of ranked data that has no ties. It is order found by using the following formula where *n* represents the number of rankings.

$$r = 1 - \frac{6(\textit{sum of the squares of the differences in the rankings})}{n(n^2 - 1)}$$

Example

My Ranking	Brian's Ranking	Difference	Difference^2
1	10	9	81
2	1	-1	1
4	8	4	16
6	3	-3	9
8	2	-6	36
7	4	-3	9
3	5	2	4
9	9	0	0
10	6	-4	16
5	7	2	4

$$r = 1 - \frac{6(81 + 1 + 16 + 9 + 36 + 9 + 4 + 16 + 4)}{10(100 - 1)} = \frac{1056}{990} = 1.067$$

Summarizing line (p. 16)   A line that follows the general pattern of a scatter plot.

Example

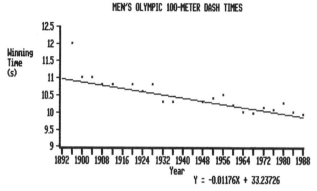

Data Insights: Scatter Plot

MEN'S OLYMPIC 100-METER DASH TIMES

Y = -0.01176X + 33.23726

T

Term (p. 45)   A number, a variable, or a *product or quotient* of numbers and variables.

Example   Some examples of terms are 2, $\frac{x}{3}$, and 7ab.

V

Visually-fit line (p. 17)   A visual estimate of a summarizing line for a set of data.

Example   A line drawn by placing a piece of string on a scatter plot and using a ruler to draw the line is a visually-fit line.

Y

y-intercept (p. 44)   The value of a function when x is 0. It is the point where the graph of an equation crosses the y-axis.

Example   The y-intercept of the equation $y = 8x + 5$ is 5.

Index

A

Area of a triangle, 37
Association, 8
   negative, 9, 14
   non-existent, 9, 14, 19
   perfect, 13
   positive, 8, 14
   strong, 10, 13
Axes, horizontal and vertical, 7

B

Box-and-whisker plots, 3

C

Circumference, 10
Cluster of points, 8
Coefficients, 45
Computer software, 11, 15
Correlation, 13
   negative, 13-14
   positive, 13
   Spearman rank order, 15
   zero, 14
Correlation coefficient, 14
Cups and counters, 53-55

D

**Data, ranked and
   unranked**, 15
Diameter, 10

F

Family of linear functions,
   43
Fitting a line to the data,
   16

G

**Grade-point average
   (GPA)**, 2
Graphing calculator, 11, 15,
   43

L

Linear equations, 36-39, 53
Linear functions, 40-48
   using to summarize data,
   49-53, 56-57
Linear model, 22
Linear patterns, 31-39
Line fitting, 16-23

M

Mathematical model, 17,
   20
Mean, 3
Mean residual, 22
Median, 3, 26
Median-fit line, 24-30, 49
   procedure for drawing,
   25-26
Median point, 26
**Models of linear
   equations,**
   53-55

O

Outliers, 9, 30

P

Paired data
   analyzing, 11-13
   displaying, 2-15
   recording, 7

Perimeter of a rectangle,
   37
Plotting points, 7

R

Rankings, 10, 15
Residual, 22
   mean, 22

S

Scatter plots, 7-10
   computer-generated, 25
   constructing, 7
   scales for, 17
   using to analyze data, 11-
   15, 22
   using to predict, 9, 16, 19,
   34
Slope, 44-48
**Spearman rank order
   correlation**, 15
Summarizing line, 16, 19
   slope of, 26

T

Term, 45
Trace function, 41

V

Variables, 8, 16
Vertical distance, 22
Visually-fit line, 17, 25

Y

y-intercept, 44

Photo Credits